Life at the Edge

Life at the Edge

. . .

READINGS FROM
SCIENTIFIC AMERICAN
MAGAZINE

Edited by

James L. Gould

Carol Grant Gould

Princeton University

W. H. FREEMAN AND COMPANY
New York

The cover
The photograph on the cover shows a fruit fly trapped by a sundew plant. The sundew is able to survive in a nitrogen-poor habitat by eating insects. Cover photograph by Thomas Eisner of Cornell University.

Some of the SCIENTIFIC AMERICAN articles in *Life at the Edge* are available as separate Offprints. For a complete list of articles now available as Offprints, write to W. H. Freeman and Company, 41 Madison Avenue, New York, New York 10010.

Library of Congress Cataloging-in-Publication Data

Life at the edge : readings from Scientific American
/ edited by James L. Gould, Carol Grant Gould.
      p.   cm.
   Bibliography: p.
   Includes index.
   ISBN 0-7167-2011-6
   1. Biology—Miscellanea.   I. Gould, James L., 1945–
II. Gould, Carol Grant.   III. Scientific American.
QH311.L645   1989
574—dc19   88-38214
                                 CIP

Printed in the United States of America

CONTENTS

SECTION III: BEATING THE ODDS

SECTION IV: THE PERILS OF PARASITISM

Note on cross-references to SCIENTIFIC AMERICAN articles: Articles included in this book are referred to by chapter number and title; articles not included in this book but available as Offprints are referred to by title, date of publication, and Offprint number; articles not in this book and not available as Offprints are referred to by title and date of publication.

Preface

Life, whether we think of the first simple replicating entities of three to four billion years ago or the millions of elaborately evolved species with which we now share the earth, depends on energy and the chemicals necessary to make use of it. The list of elements essential to life is remarkably short. Oxygen is one, of course, and carbon is the molecular backbone of every organic compound. Nitrogen is used in every amino acid and nucleotide, which are the building blocks of proteins and nucleic acids, respectively, and phosphorus is found in every nucleotide and all the molecules associated with energy transfers. A few other elements are used for metabolism in trace amounts. In terms of molecules, nearly every organism needs water, most require molecular oxygen, and all plants need carbon dioxide.

True plants use carbon dioxide and water (along with nitrogen and phosphorus from the soil) to make organic compounds and produce oxygen as a waste product. When the plant needs to use any of the energy it has stored, it uses oxygen to "burn" its fuel, generating water and carbon dioxide as by-products of that process. To take advantage of the energy stored in the plants, animals either eat the plants directly or eat other animals that do. Like the plants, they use oxygen during metabolism and produce waste water and carbon dioxide. Both plants and animals need additional water for a variety of other functions: for example, the transport of nutrients up from the roots is powered by the evaporation of water from the leaves, and animals use water to regulate temperature through evaporative cooling and to dispose of waste products. On the

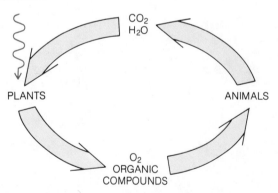

ENERGY AND NUTRIENT FLOW BETWEEN PLANTS AND ANIMALS. Plants use carbon dioxide, water and photons—as well as nitrogen and phosphorus, which are not shown—to synthesize organic compounds, producing oxygen as a by-product. Herbivores consume the plants and "burn" this food with oxygen, generating water and carbon dioxide as wastes. If the plants go ungrazed, they burn the same food chemicals (with oxygen) to grow and reproduce; predators (not shown) add an extra cycle to the right. Some simple organisms (a minute fraction of the earth's living things) are anaerobic or harvest inorganic chemical energy, and so do not fit into this cycle.

whole, therefore, the plants and animals of today are locked into an endless recycling of the same limited set of materials [see "The Cycles of Plant and Animal Nutrition," by Jules Janick, Carl H. Noller and Charles L. Rhykerd; Scientific American, September, 1976].

The ultimate source of the energy that powers all living systems is, of course, the sun. The atomic basis of the enormous energy flow that makes life work is wonderfully simple: photons add energy to the electrons with which they collide, and plants have evolved a way of preventing the energized electrons in chlorophyll from instantly reemitting light and losing that energy. Instead the energized electrons are trapped and used to charge tiny subcellular batteries. The energy thus stored in the batteries is used to activate energy-transport chemicals such as adenosine triphosphate (ATP), which can be shipped anywhere in the cell to provide the power necessary for chemical synthesis, growth, movement, or, when there is more than enough ATP, the construction of energy-storage compounds such as sugars and fats.

Burning energy-rich molecules involves a process similar to photosynthesis: the high-energy electrons stored in sugars and fats are used to charge a differ-ent set of subcellular batteries, and then the batteries are used to make ATP again. In the process the energized electrons, having been activated originally by photons seconds, days or even years earlier, lose their energy and are discarded in energy-poor carbon dioxide.

This well-oiled system of material and energy use can easily lead us to think that life is a simple, unchallenging process: electrons do their duty, energy profits are turned easily and reliably and life flourishes. This "best-of-all-possible-worlds" view, nearly universal before Darwin, ignores the billion-year struggle that led to the evolution of the first life under extremely unpromising conditions. Not only was there no air to breathe and a toxic level of radiation on the earth's surface, but every organism faced recurrent catastrophic crises, some the result of poisoning their own environments. If we assume that the plants and animals we see about us every day are the norm, we might forget that 99.9 percent of all the species the earth has ever supported are now extinct, and that for every successful species there are soon a dozen to rob it of its success through competition, predation or parasitism—if a change in the habitat does not do the job first. Then, too, our generalizations about the cycling of mate-

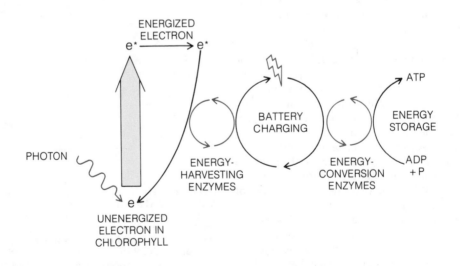

ENERGY CYCLE IN PLANTS. The energy in sunlight activates electrons, which are removed from chlorophyll before they can reemit that energy. These "excited" electrons are used to charge a membrane battery, which is used to make the energy-transfer compound adenosine triphosphate (ATP).

rials are only that. The balance is not universal, and there are habitats that are real or metaphorical deserts, short of one critical component or, just as often, toxic because of too much of another.

This book is about how life has clawed its way through a difficult birth and now manages to fill niches that, at first sight, ought to be too inhospitable for living things. We will see how fish survive in the dark and frigid waters of Antarctica while moths manage to feed, fly and mate in the snows of northern winters. We will discover how life thrives at boiling volcanic vents in the deep sea where no light penetrates for photosynthesis and the concentration of hydrogen sulfide is far above the toxic limits for the operation of the battery chargers of cellular metabolism. We will see how certain animals manage

to survive entirely on poisonous plants, and how some plants have evolved to live in nitrogen- and phosphorus-poor bogs by turning the tables and eating animals. Finally, we will enter the most dangerous and eerie world of all, that of the obligate parasite, totally dependent on its unwilling host for sustenance, avoiding or even exploiting defenses erected specifically to guard against parasitism. These examples help us appreciate not only the roles of energy, the building-block elements and the countless peripatetic activated electrons in the cycle of life, but also the blind, pragmatic and wonderfully inventive nature of natural selection itself.

James L. Gould
Carol Grant Gould

THE FIRST LIFE

. . .

Introduction

...

The evolution of early life is one of the most compelling mysteries of modern biology, but one that is slowly yielding to patient research. Certain pieces of the puzzle are clearly missing, but others seem very much in place. One of the most firmly established pieces is how the first protective coats formed in the rich organic "soup" from which life evolved.

Just as with energy flow, the key to both early and present-day membranes lies in the nature of electrons. An electron generally remains associated with a particular atom because it is negatively charged and is attracted by the positive charge of the atom's nucleus. However, not all nuclei are equally attractive: oxygen, for example, has a stronger pull (or electronegativity) than carbon or hydrogen, so that when oxygen combines with either to form a molecule—such as the H_2O of water —the shared electrons of the bond are drawn closer to the oxygen than to the nuclei of its partners. The result is that the molecules have a lopsided polarity: one side is slightly positive, while the other is marginally negative. Other molecules, being composed of atoms with similar electronegativities (carbon and hydrogen, for example), are nonpolar.

Just as electrons and nuclei attract one another because of their opposite charges, the oppositely charged ends of polar molecules bind weakly to one another. These critical attractions, known as "hydrogen bonds," hold DNA together, give proteins much of their shape and biological activity, provide water with sufficient cohesiveness to form rounded droplets on many surfaces, and cause cell membranes to form and reform. The trick behind membrane stability is that the molecules (phospholipids) involved have a dual character: one end has oxygen

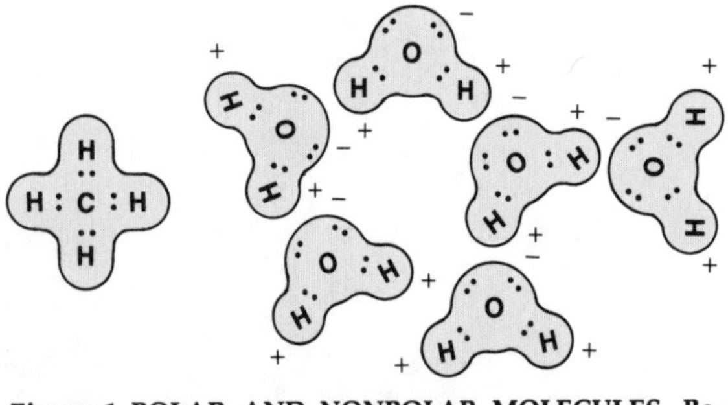

Figure 1 POLAR AND NONPOLAR MOLECULES. Because oxygen is highly electronegative, it draws the electrons it shares with other atoms closer to its own nucleus. As a result, the oxygen "corner" of a water molecule is more negative than the hydrogen corners, which are left with a slight positive charge. Because carbon and hydrogen have nearly equal electronegativities, hydrocarbons such as methane are nonpolar. (Molecular geometry can also play an important role.) Because of their slight charges, polar molecules such as water tend to form weak electrostatic associations (hydrogen bonds).

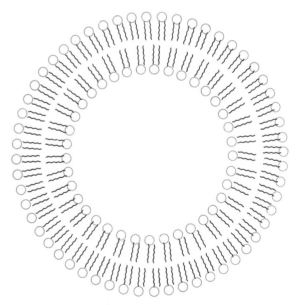

Figure 2 A LIPOSOME. The phospholipids of biological membranes have polar oxygen-rich heads (*circles*) and a pair of nonpolar hydrocarbon tails. In water the electrostatic interactions between the polar heads and water force the tails together into a clump and eventually create a bilayer sphere.

atoms and is polar, while the other, composed exclusively of carbon and hydrogen, is not. As a result, the polar ends bind electrostatically to water molecules, while the nonpolar ends tend to clump together in an electrically neutral pool, pushed together by the corralling forces between the polar ends and the surrounding water. When enough of these phospholipids are captured by the water, the pool of nonpolar ends becomes too large for stability and a two-layer sphere (a liposome) forms spontaneously, growing steadily as more membrane molecules are incorporated.

This flexible, self-repairing bilayer membrane, held in place by its polar bonds with internal and external water, creates an encapsulated milieu that can be very different chemically from the world outside. For example, small chemicals that pass in can be used to make large molecules that are unable to escape. Modern cells protect themselves from their surroundings with bilayer membranes to which specific chemical doors and pumps have been added to help control molecular in-and-out traffic [see "The Assembly of Cell Membranes," by Harvey F. Lodish and James E. Rothman; SCIENTIFIC AMERICAN, January, 1979; Offprint 1415; and "Molecules of the Cell Membrane," by Mark S. Bretscher; SCIENTIFIC AMERICAN, October, 1985]. In addition, the internal space of contemporary cells is further subdivided into special membrane-bound compartments, including the nucleus (containing the chromosomes), the mitochondria (where most metabolism occurs and which is itself further subdivided by an internal membrane), the endoplasmic reticulum (which is involved in sorting molecules), the Golgi apparatus (which also sorts) and, in plants, chloroplasts.

The best current guesses picture early liposomes forming around concentrations of energy-rich nutrients formed by the action of lightning and ionizing radiation on simple chemicals found in abundance in the earth's early atmosphere. Hydrogen cyanide, for example, is readily formed from ammonia and methane and then converted into the nucleotide adenine, which is also the backbone of ATP. Similar reactions would have produced amino acids and a host of other familiar organic compounds in pools of water, as laboratory recreations of these conditions demonstrate [see "Chemical Evolution and the Origin of Life," by Richard E. Dickerson; SCIENTIFIC AMERICAN, September, 1978]. Indeed, many meteorites and comets contain abundant inorganically formed organic compounds. Natural selection must have been at work from the outset, favoring the liposomes with the most useful chemistry, those able to concentrate the most useful building blocks efficiently and exclude any that

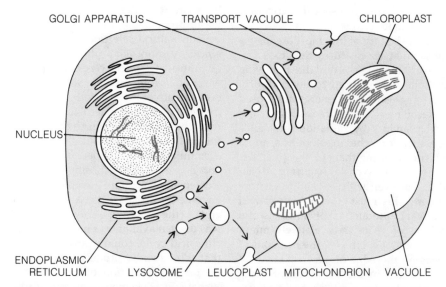

GOLGI APPARATUS   TRANSPORT VACUOLE   CHLOROPLAST

NUCLEUS

ENDOPLASMIC RETICULUM   LYSOSOME   LEUCOPLAST   MITOCHONDRION   VACUOLE

Figure 3 A SCHEMATIC EUCARYOTIC PLANT CELL. In addition to their external membrane, eucaryotes have a variety of internal bilayer-bound structures that serve to isolate chemically different compartments from one another.

might be toxic, and those able to divide spontaneously as they grow (as can those created in the lab) and even to maintain the electrical charge typical of modern cells. The development of the first enzymes and genes remains shrouded in mystery, although it now seems almost certain that the macromolecular middleman RNA originally served as both (as it still does in rare cases today). Now, however, information usually flows from the DNA to RNA, and most RNA is used to carry instructions for amino-acid sequences to the sites of protein synthesis. Essentially all enzymes are now proteins.

Once gene-directed life evolved and began to prosper, there must have been a rapid depletion of the pool of free organic nutrients. Chapter 1, "The Evolution of the Earliest Cells," by J. William Schopf, traces the development of life from that critical juncture. Faced with an early energy crisis, primitive life had to evolve ways to store energy in order to survive periods of dearth; the days of hand-to-mouth plenty were over. Chemical pathways for building energy-storage molecules like fats and sugars had to evolve, as well as pathways for harvesting that energy when needed. The oldest surviving pathway is glycolysis, during which glucose is converted into pyruvic acid, which is then fermented into lactic acid, acetaldehyde or ethanol, with a net energy gain of two ATP's for use in powering the cell. Glycolysis is inefficient, leaving much untapped energy in the products of fermentation, but it was (and for some present-day creatures is still) the only alternative to starvation.

At this point in the history of life, most organisms were autotrophs—that is, creatures that took energy or energy-rich materials from the nonliving world around them—as opposed to heterotrophs, which eat other organisms. The energy they used came from the declining stores of preexisting chemicals synthesized primarily by radiation and lightning; they were, therefore, chemoautotrophs, and a few species still survive using that strategy [see "Archaebacteria," by Carl R. Woese; SCIENTIFIC AMERICAN, June, 1981]. But even with the ability to store and later use energy, life must quickly have caught up with the inorganic production of nutrients and existed for millions of years as a relatively minor phenomenon on our globe. Life, for the first eons of its existence, was on the edge of survival.

The next critical step in the evolution of living organisms was the development of cyclic photosynthesis—cyclic because the electron energized by an incoming photon from the sun is quickly returned to the chlorophyll molecule from which it came. Cyclic photosynthesis is a membrane phenomenon: chlorophyll is embedded in a membrane along with the enzymes that steal the activated electron and harvest its energy; that energy is used to charge the membrane, and the electrostatic potential created is later employed to make ATP [see "The Photosynthetic Membrane," by Kenneth R. Miller; SCIENTIFIC AMERICAN, October, 1979; Offprint 1448; and "Molecular Mechanisms of Photosynthesis," by Douglas C. Youvan and Barry L. Marrs; SCIENTIFIC AMERICAN, June, 1987]. It takes about two photons to charge the membrane enough to make one ATP, and since photons are free, life must suddenly have been released from dependence on inorganic nutrient synthesis: with photosynthesis, there was enough ATP to generate nutrients from simple chemicals like carbon dioxide and ammonia (CO_2 and NH_4).

Although there are still bacteria that employ only cyclic photosynthesis, most photoautotrophs (including all the true plants) use the more efficient noncyclic form. In that process the electron's energy is boosted in two steps, and so much extra charging and other work is wrung from its energy that eight ATP's can be synthesized from two activated electrons—an eightfold improvement. The process is noncyclic because the electron is not returned to the chlorophyll but is handed to an energy-storage molecule instead; the missing electron is obtained by splitting water, which generates oxygen as a waste product.

Soon natural selection had so favored the high-efficiency noncyclic photosynthesizers that vast quantities of oxygen were being formed. At first this was not a problem: there were plenty of ions in the ocean ready to react with and bind up the free oxygen, and the anaerobic metabolism inherited from the earliest life must have persisted unaffected for quite some time. But slowly oxygen began to accumulate, and its appearance was both a blessing and a curse. On the positive side, the presence of oxygen allowed a fairly simple rearrangement of the enzymes of photosynthesis to harvest the waste energy in the pyruvate of glycolysis. As long as there was oxygen available, a new chemical pathway, the citric acid cycle, could charge the membrane battery that subsequently produced ATP [see "Cytochrome C and the Evolution of Energy Metabolism," by Richard E. Dickerson; SCIENTIFIC AMERICAN, March, 1980; Offprint 1464]. That process is an incredible eighteen times more efficient than fermentation.

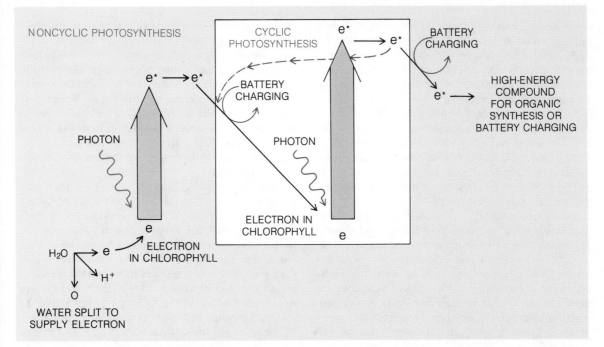

Figure 4 CYCLIC AND NONCYCLIC PHOTOSYNTHESIS. In cyclic photosynthesis (*box and dotted line*) an energized electron is removed from its chlorophyll, used to charge the membrane, and then returned (its energy spent) to the chlorophyll. In noncyclic photosynthesis, the cycle begins in a different chlorophyll and involves a second activation. The electron ends up in a multipurpose energy compound that can be used directly to power carbon fixation or to charge the membrane for subsequent ATP production. The missing electron in the first chlorophyll is replaced with one obtained by splitting water, a process that liberates oxygen.

Oxygen also made possible the evolution of high-efficiency synthetic pathways, such as the Calvin cycle, which can make energy-storage compounds quickly and cheaply. One drawback, however, was that the buildup of oxygen in the atmosphere blocked the high-energy ultraviolet rays that converted nitrogen (which organisms could not use directly) into ammonia, and therefore supplied the nitrogen needed for making proteins and nucleic acids. Suddenly there was a nitrogen crisis, and it began to pay to be a heterotroph and to obtain nitrogen by eating autotrophs. The limited availability of nitrogen must once again have placed life in jeopardy.

Necessity being the mother of evolution, natural selection intervened again, creating the expensive process of nitrogen fixation: for a cost of 30 ATP's (all the energy in a molecule of glucose) a single molecule of nitrogen can be converted into two of ammonia [see "Biological Nitrogen Fixation," by Winston J. Brill; Scientific American, March, 1977]. But that was only a temporary solution. As the concentration of oxygen continued to increase, the nitrogen-fixation pathway was poisoned: to this day fixation takes place only under nearly anaerobic conditions, and nitrogen remains a limiting nutrient (as witnessed by the flourishing fertilizer industry). Even today the colonization of barren surfaces (such as newly created volcanic islands and brick patios) would be impossible without the one group of bacteria that has evolved to have its cake and eat it too: the cyanobacteria. These noncyclic photosynthesizers grow in mats in the ocean, on rocky coasts or in other sorts of damp nitrogen deserts and fix nitrogen in special "heterocyst" cells from which oxygen is forcibly excluded. Their very success, however, sows the seeds of destruction: with a thin film of nitrogen-rich organisms to graze on, higher forms of life move in to live on the nutritious substrate created by the cyanobacteria and kill them in process.

Eventually the increasing buildup of oxygen (and ozone, which is produced when oxygens combine using the energy provided by ultraviolet light) screened the earth's surface from enough deadly radiation that life could finally emerge from the protection of the water and evolve on land. On balance oxygen was a net benefit, but nearly every kind of life that existed before the oxygen revolution perished as a result.

Another consequence of the appearance of abundant oxygen was the evolution of higher cells. Life consists of bacteria, which have a single membrane, and eukaryotes ("true cells"), which have membrane-bound internal structures. Eukaryotes evolved from synergistic associations of bacteria; the mitochondria of present-day cells descended from heterotrophs that depended on oxygen and battery charging to make their living; and chloroplasts were once free-living noncyclic photosynthesizers whose modern descendents are the cyanobacteria. Although the associations that form the symbiotic corporations we know as eukaryotes are now obligate—that is, mitochondria and chloroplasts cannot survive on their own since too many of their genes have been transferred to the cell nucleus—there are still many examples of intermediate stages of cooperation. Perhaps the most common are lichens, which are symbiotic associations between fungi and either algae or photosynthetic bacteria (depending on the species). With the evolution of eukaryotes came the more complex genetic mechanisms (mitosis, meiosis, genetic recombination), possible now with the development of a true nucleus, tricks that help organisms adapt to changing conditions while protecting the integrity of the genetic message.

Once the complex multicellular eukaryotes began to appear, the first large specialized predators evolved. As Chapter 2, "The Emergence of Animals," shows us, the very success of eukaryotes led to the development of parasites and predators that fed upon the autotrophs—the producers—and then parasites and predators that focused on other predators. Success creates new niches, and we are reminded that as a result change is inevitable, as selection dictates ploy and counterploy in the three-billion-year-old life-and-death struggle to survive.

The Evolution of the Earliest Cells

For some three billion years the only living things were primitive microorganisms. These early cells gave rise to biochemical systems and the oxygen-enriched atmosphere on which modern life depends.

. . .

J. William Schopf

September, 1978

When *On the Origin of Species* appeared in 1859, the history of life could be traced back to the beginning of the Cambrian period of geologic time, to the earliest recognized fossils, forms that are now known to have lived more than 500 million years ago. A far longer prehistory of life has since been discovered: it extends back through geologic time almost three billion years more. During most of that long Precambrian interval the only inhabitants of the earth were simple microscopic organisms, many of them comparable in size and complexity to modern bacteria. The conditions under which these organisms lived differed greatly from those prevailing today, but the mechanisms of evolution were the same. Genetic variations made some individuals better fitted than others to survive and to reproduce in a given environment, and so the heritable traits of the better-adapted organisms were more often represented in succeeding generations. The emergence of new forms of life through this principle of natural selection worked great changes in turn on the physical environment, thereby altering the conditions of evolution.

One momentous event in Precambrian evolution was the development of the biochemical apparatus of oxygen-generating photosynthesis. Oxygen released as a by-product of photosynthesis accumulated in the atmosphere and effected a new cycle of biological adaptation. The first organisms to evolve in response to this environmental change could merely tolerate oxygen; later cells could actively employ oxygen in metabolism and were thereby enabled to extract more energy from foodstuff.

A second important episode in Precambrian history led to the emergence of a new kind of cell, in which the genetic material is aggregated in a distinct nucleus and is bounded by a membrane. Such nucleated cells are more highly organized than those without nuclei. What is most important, only nucleated cells are capable of advanced sexual reproduction, the process whereby the genetic variations of the parents can be passed on to the offspring in new combinations. Because sexual reproduction allows novel adaptations to spread quickly through a population its development accelerated the pace of evolutionary change. The large, complex, multicellular forms of life that have appeared and quickly diversified since the beginning of the Cambrian period are without exception made up of nucleated cells.

The boundary between the Precambrian era and the Cambrian period has traditionally been viewed as a sharp discontinuity. In Cambrian strata there are abundant fossils of marine plants and animals: seaweeds, worms, sponges, mollusks, lampshells and, what are perhaps most characteristic of the period, the early arthropods called trilobites. It was thought for many years that fossils were entirely absent in the underlying Precambrian strata. The Cambrian fauna seemed to come into existence abruptly and without known predecessors.

Life could not have begun with organisms as complex as trilobites. In *On the Origin of Species* Darwin wrote: "To the question why we do not find rich fossiliferous deposits belonging to . . . periods prior to the Cambrian system, I can give no satisfactory answer. . . . The case at present must remain inexplicable; and may be truly urged as a valid argument against the views here entertained." The argument is no longer valid, but it is only in the past 20 years or so that a definitive answer to it has been found.

One part of the answer lies in the discovery of primitive fossil animals in rocks below the earliest Cambrian strata. The fossils include the remains of jellyfishes, various kinds of worms and possibly sponges, and they make up a fauna quite distinct from that of the predominantly shelled animals of the Cambrian period. These discoveries, however, extend the fossil record by only about 100 million years, less than four percent of the Precambrian era. It can still be asked: What came before?

Since the 1950's a far-reaching explanation has emerged. It has come to be recognized that not only are many Precambrian rocks fossil-bearing but also Precambrian fossils can be found even in some of the most ancient sedimentary deposits known. These fossils had escaped notice earlier largely because they are the remains only of microscopic forms of life. They exist in structures called stromatolites, from the Greek *stroma*, meaning bed or coverlet, and *lithos*, meaning stone.

A surprising amount of information can be derived from the fossil remains of a microorganism. Size, shape and degree of morphological complexity are among the most easily recognized features, but under favorable circumstances even details of the internal structure of cells can be discerned. In re-

Figure 5 LIVING STROMATOLITES were photographed at Shark Bay in Australia. Elsewhere stromatolites are rare because of grazing by invertebrates. Here the invertebrates are excluded because the water is too salty for them; in the Precambrian era they had not yet evolved. In size and form the modern stromatolites are much like the fossil structures, and they are produced by the growth of cyanobacteria and other prokaryotes in matlike communities. The discovery of such living stromatolites has confirmed the biological origin of the fossil ones.

tracing the course of Precambrian evolution, however, there is no need to rely exclusively on the fossil record. An entirely independent archive has been preserved in the metabolism and the biochemical pathways of modern, living cells. No living organism is biochemically identical with its Precambrian antecedents, but vestiges of earlier biochemistries have been retained. By studying their distribution in modern forms of life it is sometimes possible to deduce when certain biochemical capabilities first appeared in the evolutionary sequence.

Still another independent source of information about the early evolutionary progression is based neither on living nor on fossil organisms but on the inorganic geological record. The nature of the minerals found there reflects physical conditions at the time the minerals were deposited, conditions that may have been influenced by biological innovations. In order to understand the introduction of oxygen into the early atmosphere, for example, all three fields of study must be called on to testify. The mineral record tells when the change took place, the fossil record reveals the organisms responsible and the distribution of biochemical capabilities among modern organisms puts the development in its proper evolutionary context.

Since the 1960s it has become apparent that the greatest division among living organisms is not between plants and animals but between organisms whose cells have nuclei and those that lack nuclei. In terms of biochemistry, metabolism, genetics and intracellular organization, plants and animals are very similar; all such higher organisms, however, are quite different in these features from bacteria and the so-called blue-green algae, the principal types of non-nucleated life. Recognition of this discontinuity has been important for understanding the early stages of biological history.

Organisms whose cells have nuclei are called eukaryotes, from the Greek roots *eu-*, meaning well or true, and *karyon,* meaning kernel or nut. Cells without nuclei are prokaryotes, the prefix *pro-* meaning before. All green plants and all animals are eukaryotes. So are the fungi, including the molds and the yeasts, and protists such as *Paramecium* and *Euglena.* The prokaryotes include only two groups of organisms, the bacteria and the blue-green algae. The latter produce oxygen through photosynthesis like other algae and higher plants, but they are now known to be true bacteria. I shall therefore refer to blue-green algae by an alternative and more descriptive name, the cyanobacteria.

Several important traits distinguish eukaryotes from prokaryotes. In the nucleus of a eukaryotic cell the DNA is organized in chromosomes and is enclosed by an intracellular membrane; many prokaryotes have only a single loop of DNA, which is loose in the cytoplasm of the cell. Prokaryotes reproduce asexually by the comparatively simple process of binary fission. In contrast, asexual reproduction in eukaryotic cells takes place through the complicated process of mitosis, and most eukaryotes can also reproduce sexually through meiosis and the subsequent fusion of sex cells. (The ''parasexual'' reproduction of some prokaryotes differs markedly from advanced eukaryotic sexuality.) Eukaryotic

cells are generally larger than prokaryotic ones, although the range of sizes overlaps, and almost all prokaryotes are unicellular organisms whereas the majority of eukaryotes are large, complex and many-celled. A mammalian animal, for example, can be made up of billions of cells, which are highly differentiated in both structure and function.

An intriguing feature of eukaryotic cells is that they have within them smaller membrane-bounded subunits, or organelles, the most notable being mitochondria and chloroplasts. Mitochondria are present in all eukaryotes, where they play a central role in the energy economy of the cell. Chloroplasts are present in some protists and in all green plants and are responsible for the photosynthetic activities of those organisms. It has been suggested that both mitochondria and chloroplasts may be evolutionary derivatives of what were once free-living microorganisms, an idea discussed in particular by Lynn Margulis of Boston University. The modern chloroplast, for example, may be derived from a cyanobacterium that was engulfed by another cell and that later established a symbiotic relationship with it. In support of this hypothesis it has been noted that both mitochondria and chloroplasts contain a small fragment of DNA whose organization is somewhat like that of prokaryotic DNA. In the past several years the testing of this hypothesis has generated a large body of data on the comparative biochemistry of modern microorganisms, data that also provide clues to the evolution of life in the Precambrian.

One further difference between prokaryotes and eukaryotes is of particular importance in the study of their evolution: the extent to which the two types of organisms tolerate oxygen. Among the prokaryotes oxygen requirements are quite variable. Some bacteria cannot grow or reproduce in the presence of oxygen; they are classified as obligate anaerobes. Others can tolerate oxygen but can also survive in its absence; they are facultative anaerobes. There are also prokaryotes that grow best in the presence of oxygen but only at low concentrations, far below that of the present atmosphere. Finally, there are fully aerobic prokaryotes, forms that cannot survive without oxygen.

In contrast to this variety of adaptations the eukaryotes present a pattern of great consistency: with very few exceptions they have an absolute requirement for oxygen, and even the exceptions seem to

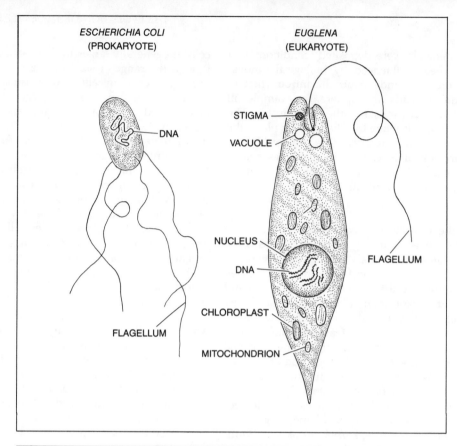

	PROKARYOTES	EUKARYOTES
ORGANISMS REPRESENTED	BACTERIA AND CYANOBACTERIA	PROTISTS, FUNGI, PLANTS AND ANIMALS
CELL SIZE	SMALL, GENERALLY 10 TO 100 MICROMETERS	LARGE, GENERALLY 1 TO 10 MICROMETERS
METABOLISM AND PHOTOSYNTHESIS	ANAEROBIC OR AEROBIC	AEROBIC
MOTILITY	NONMOTILE OR WITH FLAGELLA MADE OF THE PROTEIN FLAGELLIN	USUALLY MOTILE, CILIA OR FLAGELLA CONSTRUCTED OF MICROTUBULES
CELL WALLS	OF CHARACTERISTIC SUGARS AND PEPTIDES	OF CELLULOSE OR CHITIN, BUT LACKING IN ANIMALS
ORGANELLES	NO MEMBRANE-BOUNDED ORGANELLES	MITOCHONDRIA AND CHLOROPLASTS
GENETIC ORGANIZATION	LOOP OF DNA IN CYTOPLASM	DNA ORGANIZED IN CHROMO-SOMES AND BOUNDED BY NUCLEAR MEMBRANE
REPRODUCTION	BY BINARY FISSION	BY MITOSIS OR MEIOSIS
CELLULAR ORGANIZATION	MAINLY UNICELLULAR	MAINLY MULTICELLULAR, WITH DIFFERENTIATION OF CELLS

be evolutionary derivatives of oxygen-dependent organisms. This observation leads to a simple hypothesis: the prokaryotes evolved during a period when environmental oxygen concentrations were changing, but by the time the eukaryotes arose the oxygen content was stable and relatively high.

One indication that eukaryotic cells have always been aerobic is provided by mitotic cell division, a process that can be considered a definitive characteristic of the group. Many eukaryotic cells can survive temporary deprivation of oxygen and can even carry on some metabolic functions; it appears that no cell, however, can undergo mitosis unless oxygen is available at least in low concentration.

The pathways of metabolism itself—the biochemical mechanisms by which an organism extracts energy from foodstuff—provide more detailed evidence. In eukaryotes the central metabolic process is respiration, which in overall terms can be described as the burning of the sugar glucose with oxygen to yield carbon dioxide, water and energy. Some prokaryotes (the aerobic or facultative ones) are also capable of respiration, but many derive their energy solely from the simpler process of fermentation. In bacterial fermentation glucose is not combined with oxygen (or with any other substance from outside the cell) but is simply broken down into smaller molecules. In both respiration and fermentation part of the energy released through the decomposition of glucose is captured in the form of high-energy phosphate bonds, usually in molecules of adenosine triphosphate (ATP). The rest of the energy is lost from the cell as heat.

Respiratory metabolism has two main components: a short series of chemical reactions, collectively called glycolysis, and a longer series called the citric acid cycle. In glycolysis a glucose molecule, with six carbon atoms, is broken down into two molecules of pyruvate, each having three carbon

Figure 6 GREATEST DIVISION among organisms is the one separating cells with nuclei (eukaryotes) from those without nuclei (prokaryotes). The only prokaryotes are bacteria and cyanobacteria, and here they are represented by the bacterium *Escherichia coli* (*top left*). All other organisms are eukaryotes, including higher plants and animals, fungi and protists such as *Euglena* (*top right*). Eukaryotic cells are by far the more complex ones, and some of the organelles they contain, such as mitochondria and chloroplasts, may be derived from prokaryotes that established a symbiotic relationship with the host cell. Prokaryotes vary widely in their tolerance of or requirement for free oxygen. All eukaryotes require oxygen for metabolism and for the synthesis of various substances.

atoms. No oxygen is required for glycolysis, but on the other hand it releases only a little energy with a net gain of only two molecules of ATP.

The fuel for the citric acid cycle is the pyruvate formed by glycolysis. Through a series of enzyme-controlled reactions the carbon atoms of the pyruvate are oxidized and the oxidations are coupled to other reactions that result in the synthesis of ATP. For each two molecules of pyruvate (and hence for each molecule of glucose entering the sequence) 34 additional molecules of ATP are formed. The complete respiratory pathway is thus far more effective than glycolysis alone. By breaking down the glucose to simple inorganic molecules (carbon dioxide and water) respiration liberates virtually all the biologically usable energy stored in the chemical bonds of the sugar.

The metabolism of the prokaryotes immediately suggests an evolutionary relationship between them and the eukaryotes: up to a point fermentation is indistinguishable from glycolysis. In bacterial fermentation a molecule of glucose is split into two molecules of pyruvate, with a net yield of two molecules of ATP. As in glycolysis, no oxygen is required for the process. In anaerobic prokaryotes, however, the metabolic pathway essentially ends at pyruvate. The only further reactions transform the pyruvate into such compounds as lactic acid, ethyl alcohol, or carbon dioxide, which are excreted by the cell as wastes.

The similarity of fermentation in prokaryotes to glycolysis in eukaryotes seems too close to be a coincidence, and the assumption of an evolutionary relationship between the two groups provides a ready explanation. It seems likely that anaerobic fermentation became established as an energy-yielding process early in the history of life. When atmospheric oxygen became available for metabolism, it offered the potential for extracting 18 times as much useful energy from carbohydrate: a net yield of 36 molecules of ATP instead of only two molecules. The oxygen-dependent reactions did not, however, simply replace the anaerobic ones; they were appended to the existing anaerobic pathway.

Further evidence for this proposed evolutionary sequence can be found in the behavior of some eukaryotic cells under conditions of oxygen deprivation. In mammalian muscle cells, for example, prolonged exertion can demand more oxygen than the lungs and the blood can supply. The citric acid cycle is then disabled, but the cells continue to func-

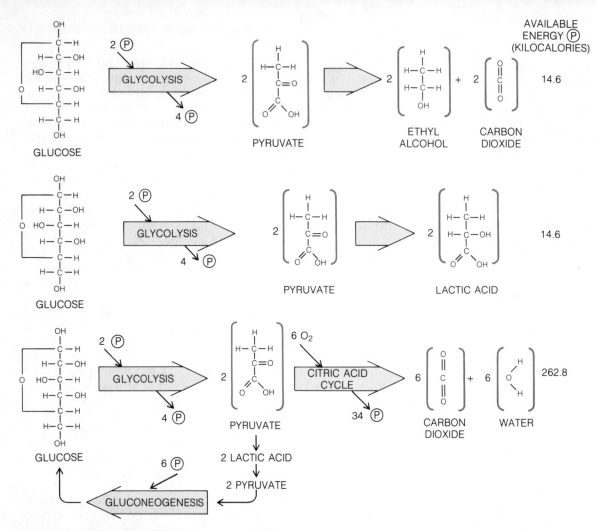

Figure 7 METABOLIC PATHWAYS by which cells extract energy from foodstuff apparently evolved in response to an increase in free oxygen. In anaerobic organisms (those that live without oxygen) glucose is broken down through fermentation: each molecule of glucose is split into two molecules of pyruvate, the process called glycolysis, with a net gain of two phosphate bonds. In bacterial fermentation the pyruvate is converted into products such as lactic acid or ethyl alcohol and carbon dioxide, which are excreted as wastes. The metabolic system of aerobic organisms (those that require oxygen) is respiration. It begins with glycolysis, but the pyruvate is treated not as a waste but as a substrate for a further series of reactions that make up the citric acid cycle. Respiration releases far more energy than fermentation; as a result 36 phosphate bonds are formed instead of two.

tion, albeit at reduced efficiency, through glycolysis alone. Under such conditions of oxygen debt pyruvate is not consumed in the cell, but in the liver it can be converted back into glucose (at a cost in energy of six ATP molecules). Significantly the pyruvate itself is not transported to the liver but instead is converted into lactic acid, which in the liver must then be returned to the form of pyruvate. This use of lactic acid may represent a vestige of an earlier, bacterial pathway that under aerobic condi-

tions has been suppressed. Indeed, the oxygen-starved muscle cell seems to revert to a more primitive, entirely anaerobic form of metabolism.

The development of an oxygen-dependent biochemistry can also be traced through a consideration of reaction sequences in the synthesis of various biological molecules. Once again stages in the synthetic pathway that emerged early in the Precambrian can be expected to proceed in the absence of oxygen. Reaction steps nearer the end product of the pathway, which were presumably added at a later stage, might with increasing frequency require oxygen. The distribution of the oxygen-demanding steps among various kinds of organisms could also have evolutionary significance. If only one pathway has evolved for the synthesis of a class of biochemical substances, then primitive forms of life might be expected to exhibit only the initial, anaerobic steps. Organisms that arose later might exhibit progressively longer, oxygen-dependent synthetic sequences.

Indeed, some syntheses have an intrinsic requirement for oxygen. Molecular oxygen is needed, for example, in the synthesis of bile pigments in vertebrates, of chlorophyll *a* in higher plants and of the amino acids hydroxyproline and, in animals, tyrosine. The oxygen dependence of two synthetic pathways in particular has been determined in detail. One of these pathways controls the manufacture of a class of compounds that includes the sterols and the carotenoids and the other is concerned with the synthesis of fatty acids.

Two observations about the evolution of these biosynthetic pathways are appropriate. Even in groups of organisms that have long been aerobic the first steps in the synthesis are independent of the oxygen supply; molecular oxygen enters the reaction sequence only at later stages. In a similar way the most primitive living organisms, the anaerobic bacteria, are capable only of the first segments of the pathway, the anaerobic segments. The more complex aerobic bacteria and the photosynthetic cyanobacteria have longer synthetic pathways, including some steps that require oxygen. Advanced eukaryotes, such as vertebrate animals and higher plants have long, branched synthetic pathways, with many steps in which molecular oxygen is required.

Comparisons of the metabolism and biochemistry of prokaryotes and eukaryotes thus provide strong evidence that the latter group arose only after a substantial quantity of oxygen had accumulated in the atmosphere. Hence it is of interest to ask when eukaryotic cells first appeared. It seems apparent that an oxygen-rich atmosphere cannot have developed later than this signal evolutionary event.

The primary means of assigning a date to the origin of the eukaryotes is through the fossil record. Because this field of study is so new, however, the available information is scanty and often difficult to interpret. It is rarely a straightforward task to identify a microscopic, single-cell organism as being eukaryotic merely from an examination of its fossilized remains. And even when a fossil has been identified as unequivocally eukaryotic the available radioactive-isotope methods of dating can rarely assign it a precise age. At best such methods have an accuracy of only about plus or minus 5 percent. What is more, the age determinations are generally carried out on rocks that were once molten, such as volcanic lavas, whereas the fossils are found in sedimentary deposits. Consequently the stratum of the fossil itself usually cannot be dated; it is merely assigned an age somewhere between the ages of the nearest underlying and overlying datable rock units.

In spite of these difficulties there is now substantial evidence for the existence of eukaryotic fossils in rocks hundreds of millions of years older than the earliest Phanerozoic strata. The evidence is of two kinds: microfossils that display a morphological or organizational complexity judged to be of eukaryotic character, and the presence of fossil cells whose size is typical only of eukaryotes.

The evidence from relatively complex microscopic fossils includes the following: (1) branched filaments, made up of cells with distinct cross walls and resembling modern fungi or green algae, from the Olkhin formation of Siberia, a deposit thought to be about 725 million years old (but with a known age of between 680 and 800 million years); (2) complex, flask-shaped microfossils from the Kwagunt formation in the eastern Grand Canyon, thought to be about 800 (or 650 to 1,150) million years old; (3) fossils of unicellular algae containing intracellular membranes and small, dense bodies that may represent preserved organelles, from the Bitter Springs formation of central Australia, dated at approximately 850 (or 740 to 950) million years; (4) a group of four sporelike cells in a tetrahedral configuration that may have been produced by mitosis or possibly meiosis, also from Bitter Springs rocks; (5) spiny cells or algal cysts several hundred micrometers in diameter and with unquestionable affinities to eukaryotic organisms, from Siberian shales that are reportedly 950 (or 750 to 1,050) million years old; (6) highly branched filaments of large diameter and with rare cross walls, similar in some respects to certain green or golden-green eukaryotic algae, from the Beck Spring dolomite of southeastern California (1,300, or 1,200 to 1,400 million years old) and from the Skillogalee dolomite of South Australia (850, or 740 to 867 million years old); (7) sphe-

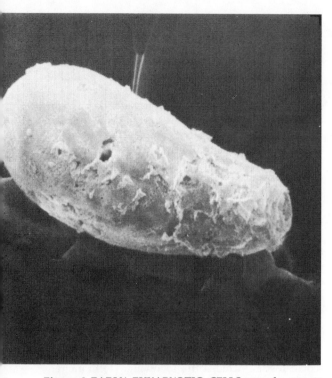
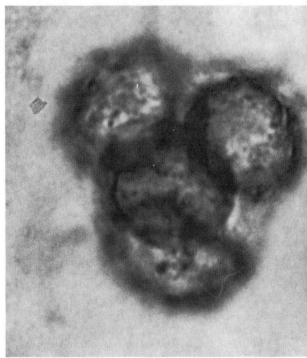

Figure 8 EARLY EUKARYOTIC CELLS may be represented among Precambrian microfossils. The gourd-shaped cell at the left, from shales in the Grand Canyon thought to be 800 million years old, is morphologically more complex than any known prokaryote; it is also larger, about 100 micrometers long. The cluster of cells shown at the right is from sediments in central Australia thought to be 850 million years old. The cells are only 10 micrometers across, but their tetrahedral arrangement suggests they formed as a result of mitosis or possibly meiosis, mechanisms of cell division known only in eukaryotes.

roidal microfossils described as exhibiting two-layered walls and having "medial splits" on their surface, and which may represent an encystment stage of a eukaryotic alga, from shales 1,400 (or 1,280 to 1,450) million years old in the McMinn formation of northern Australia; (8) a tetrahedral group of four small cells, resembling spores produced by mitotic cell division of some green algae, from the Amelia dolomite of northern Australia, approaching 1,500 (or 1,390 to 1,575) million years in age; (9) unicellular fossils that appear to be exceptionally well preserved and that are reported to contain small membrane-bounded structures that could be remnants of organelles, from the Bungle-Bungle dolomite in the same region as the Amelia dolomite and of approximately the same age.

Thus the earliest of these eukaryote-like fossils are probably somewhat less than 1,500 million years old. Numerous types of microfossils have been discovered in older sediments, but none of them seems to be a strong candidate for identifica-

tion as eukaryotic. For example, the well-studied Canadian fossils of the Gunflint and Belcher Island iron formations, which are about two billion years old, have been interpreted as exclusively prokaryotic. The testimony of these as yet rare and unusual specimens can be checked through statistical studies of the sizes of known Precambrian microfossils. The size ranges of prokaryotes and eukaryotes overlap, so that a particular fossil cannot always be classified unambiguously on the basis of size alone; by cataloguing the measured sizes in a large sample of fossils, however, it may be possible to determine whether or not eukaryotic cells are present.

Studies of the size of unicellular fossils suggest that there is a break in the fossil record between 1,400 and 1,500 million years ago. Below this horizon cells with eukaryotelike traits are rare or absent; above it they become increasingly common. Moreover, the data suggest that the diversification of the eukaryotes began shortly after the cell type first appeared, apparently within the next few hundred

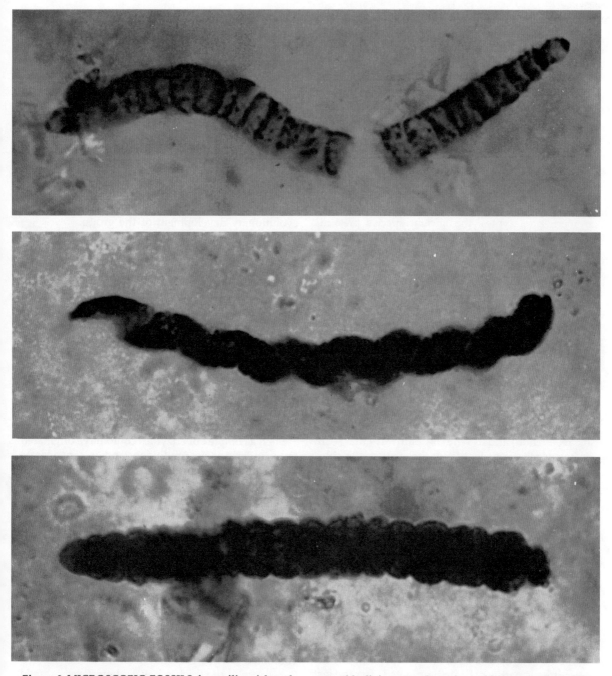

Figure 9 MICROSCOPIC FOSSILS from silica-rich rocks in the Bitter Springs formation of central Australia, deposited about 850 million years ago, or late in the Precambrian era. The rocks have the layered structure of stromatolites. Among Precambrian fossils these specimens are exceptionally well preserved; their petrified cell walls are composed of organic matter and have retained their three-dimensional form. In size, structure and ecological setting they resemble living cyanobacteria, or blue-green algae. Like their modern counterparts, the fossil forms were presumably capable of photosynthesis, and similar cyanobacteria some billion years earlier were evidently responsible for the first rapid release of oxygen into the earth's atmosphere. Organisms in these photomicrographs are about 60 micrometers long.

million years. By a billion years ago there had been substantial increases in cell size, in morphological complexity and in the diversity of species. All these indicators also suggest, of course, that oxygen-dependent metabolism, which is highly developed even in the most primitive eukaryotes, had already become established by about 1.5 billion years ago.

The prokaryotes that must have held exclusive sway over the earth before the development of eukaryotic cells were less diverse in form, but they were probably more varied in metabolism and biochemistry than their eukaryotic descendants. Like modern prokaryotes, the ancient species presumably varied over a broad range in their tolerance of oxygen, all the way from complete intolerance to absolute need. In this regard one group of prokaryotes, the cyanobacteria, are of particular interest in that they were largely responsible for the development of an oxygen-rich atmosphere.

Like higher plants, cyanobacteria carry out aerobic photosynthesis, a process that in overall effect (although not in mechanism) is the reverse of respiration. The energy of sunlight is employed to make carbohydrates from water and carbon dioxide, and molecular oxygen is released as a by-product. The cyanobacteria can tolerate the oxygen they produce and can make use of it both metabolically (in aerobic respiration) and in synthetic pathways that seem to be oxygen dependent (as in the synthesis of chlorophyll a). Nevertheless, the biochemistry of the cyanobacteria differs from that of green, eukaryotic plants and suggests that the group originated during a time of fluctuating oxygen concentration. For example, although many cyanobacteria can make unsaturated fatty acids by oxidative desaturation, some of them can also employ the anaerobic mechanism of adding a double bond during the elongation of the chain. In a similar manner oxygen-dependent syntheses of certain sterols can be carried out by some cyanobacteria, but the amounts of the sterols made in this way are minuscule compared with the amounts typical of eukaryotes. In other cyanobacteria those sterols are not found at all, the biosynthetic pathway being terminated after the last anaerobic step. Hence in their biochemistry the cyanobacteria seem to occupy a middle ground between the anaerobes and the eukaryotes.

In metabolism too the cyanobacteria occupy an intermediate position. They flourish today in fully oxygenated environments, but physiological experiments indicate that for many species optimum growth is obtained at an oxygen concentration of about 10 percent, which is only half that of the present atmosphere. Both photosynthesis and respiration are increasingly inhibited when the oxygen concentration exceeds that optimum level. It has recently been discovered that some cyanobacteria can switch the cellular machinery of aerobic metabolism on and off according to the availability of oxygen. Under anoxic conditions these species not only halt respiration but also adopt an anaerobic mode of photosynthesis, employing hydrogen sulfide (H_2S) instead of water and releasing sulfur instead of oxygen. This capability for anaerobic metabolism is probably a relic of an earlier stage in the evolutionary development of the group.

Another activity of some cyanobacteria that seems to reflect an earlier adaptation to anoxic conditions is nitrogen fixation. Nitrogen is an essential element of life, but it is biologically useful only in "fixed" form, for example combined with hydrogen in ammonia (NH_3). Only prokaryotes are capable of fixing nitrogen (although they often do so in symbiotic relationships with higher plants). The crucial complex of enzymes for fixation, the nitrogenases, is highly sensitive to oxygen. In cell-free extracts nitrogenases are partially inhibited by as little as .1 percent of free oxygen, and they are irreversibly inactivated in minutes by exposure to oxygen concentrations of only about 5 percent.

Such a complex of enzymes could have originated only under anoxic conditions, and it can operate today only if it is protected from exposure to the atmosphere. Many nitrogen-fixing bacteria provide that protection simply by adopting an anaerobic habitat, but among the cyanobacteria a different strategy has developed: the nitrogenase enzymes are protected in specialized cells, called heterocysts, whose internal milieu is anoxic. The heterocysts lack certain pigments essential for photosynthesis, and so they generate no oxygen of their own. They have thick cell walls and are surrounded by a mucilaginous envelope that retards the diffusion of oxygen into the cell. Finally, they are equipped with respiratory enzymes that quickly consume any uncombined oxygen that may leak in.

Because of the thick cell walls heterocysts should be comparatively easy to recognize in fossil material. Indeed, possible heterocysts have been reported from several Precambrian rock units, the oldest being about 2.2 billion years in age. If these cells are indeed heterocysts, they may be taken as a sign that free oxygen was present by then, at least in small concentrations.

Nitrogen fixation has a high cost in energy, and the capability for it would therefore seem to confer a selective advantage only when fixed nitrogen is a scarce resource. Today the main sources of fixed nitrogen are biological and industrial, but biologically usable nitrate (NO_3^-) is formed by the reaction of atmospheric nitrogen and oxygen. In the anoxic atmosphere of the early Precambrian the latter mechanism would obviously have been impossible. The lack of atmospheric oxygen would also have indirectly reduced the concentration of ammonia to very low levels. Ammonia is dissociated into nitrogen and hydrogen by ultraviolet radiation, most of which is filtered out today by a layer of ozone (O_3) high in the atmosphere; without free oxygen there would have been little ozone, and without this protective shield atmospheric ammonia would have been quickly destroyed.

It is likely that the capability for nitrogen fixation developed early in the Precambrian among primitive prokaryotic organisms and in an environment where fixed nitrogen was in short supply. The vulnerability of the nitrogenase enzymes to oxidation was of no consequence then, since the atmosphere had little oxygen. Later, as the photosynthetic activities of the cyanobacteria led to an increase in atmospheric oxygen, some nitrogen fixers adopted an anaerobic habitat and others developed heterocysts. By the time eukaryotes appeared, apparently more than half a billion years later, oxygen was abundant and fixed nitrogen (both NH_3 and NO_3^-) was probably less scarce, and so the eukaryotes never developed the enzymes needed for nitrogen fixation.

At present oxygen-releasing photosynthesis by green plants, cyanobacteria and some protists is responsible for the synthesis of most of the world's organic matter. It is not, however, the only

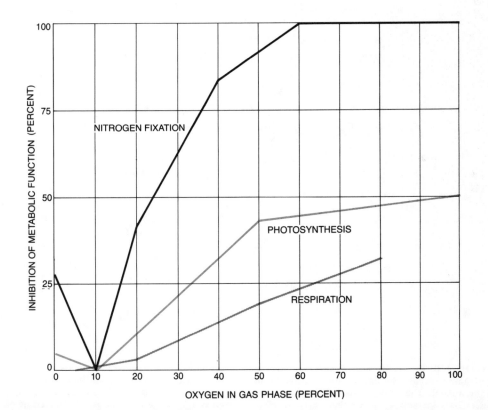

Figure 10 OXYGEN INHIBITION of metabolic functions in cyanobacteria suggests that these aerobic prokaryotes are adapted to an optimum oxygen concentration of about 10 percent, or roughly half the oxygen concentration of the earth's present atmosphere. Nitrogen fixation is completely halted by high oxygen levels, but even respiration, which requires oxygen, can be partially inhibited. Data are for the heterocyst-forming cyanobacterium *Anabaena flos-aquae*.

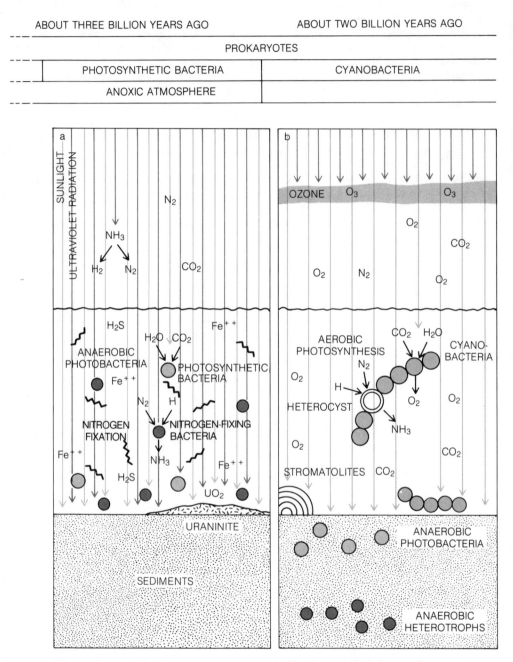

ABOUT THREE BILLION YEARS AGO   ABOUT TWO BILLION YEARS AGO

PROKARYOTES

PHOTOSYNTHETIC BACTERIA   CYANOBACTERIA

ANOXIC ATMOSPHERE

Figure 11 THE FIRST LIVING CELLS survived by fermenting organic molecules formed nonbiologically in the anoxic environment. The first photosynthetic organisms were also entirely anaerobic (a). Another early development was nitrogen fixation. A little more than two billion years ago aerobic photosynthesis began. Oxygen was generated until, after some 100 million years, it reacted with iron dissolved in the oceans. Only when the oceans had been swept free of iron and similar materials did the concentrations of free oxygen begin to rise (b). Anaerobic orga-

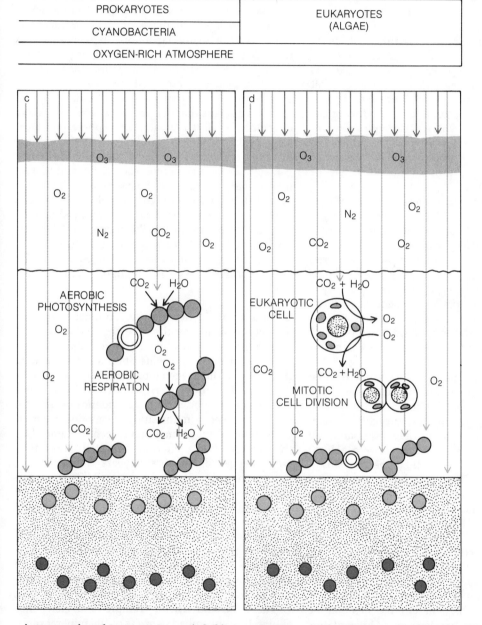

ABOUT 1.5 BILLION YEARS AGO

PROKARYOTES	EUKARYOTES
CYANOBACTERIA	(ALGAE)
OXYGEN-RICH ATMOSPHERE	

nisms were forced to retreat to anoxic habitats or develop protective heterocyst cells. Atmospheric oxygen also created a layer of ozone (O_3) that filtered out most ultraviolet radiation. Cells evolved that could employ oxygen in respiration (c). The result was a great improvement in metabolic efficiency. About 1,450 million years ago, the first eukaryotic cells emerged (d).

mechanism of photosynthesis. The alternative systems are confined to a few groups of bacteria that on a global scale seem to be of minor importance today but that may have been far more significant in the geological past.

The several groups of photosynthetic bacteria differ from one another in their pigmentation, but they are alike in one important respect: unlike the photosynthesis of cyanobacteria and eukaryotes, all bacterial photosynthesis is a totally anaerobic process. Oxygen is not given off as a by-product of the reaction, and the photosynthesis cannot proceed in the presence of oxygen. Whereas oxygen appears to be required in green plants for the synthesis of chlorophyll *a*, oxygen inhibits the synthesis of bacteriochlorophylls.

The anaerobic nature of bacterial photosynthesis seems to present a paradox: photosynthetic organisms thrive where light is abundant, but such environments are also generally ones having high concentrations of oxygen, which poisons bacterial photosynthesis. These contradictory needs can be explained if it is assumed that anaerobic photosynthesis evolved among primitive bacteria early in the Precambrian, when the atmosphere was essentially anoxic. The photosynthesizers could thus have lived in matlike communities in shallow water and in full sunlight.

Somewhat later such bacteria gave rise to the first organisms capable of aerobic photosynthesis, the precursors of modern cyanobacteria. For the anaerobic photosynthetic bacteria the molecular oxygen released by this mutant strain was a toxin, and as a result the aerobic photosynthesizers were able to supplant the anaerobic ones in the upper portions of the mat communities. The anaerobic species became adapted to the lower parts of the mat, where there is less light but also a lower concentration of oxygen. Many photosynthetic bacteria occupy such habitats today.

Photosynthetic bacteria were surely not the first living organisms, but the history of life in the period that preceded their appearance is still obscure. What little information can be inferred about that early period, however, is consistent with the idea that the environment was then largely anoxic. One tentative line of evidence rests on the assumption that among organisms living today those that are simplest in structure and in biochemistry are probably the most closely related to the earliest forms of life. Those simplest organisms are bacteria of the clostridial and methanogenic types, and they are all obligate anaerobes.

There is even a basis for arguing that anoxic conditions must have prevailed during the time when life first emerged on the earth. The argument is based on the many laboratory experiments that have demonstrated the synthesis of organic compounds under conditions simulating those of the primitive planet. These syntheses are inhibited by even small concentrations of molecular oxygen. Hence it appears that life probably would not have developed at all if the early atmosphere had been oxygen-rich. It is also significant that the starting materials for such experiments often include hydrogen sulfide and carbon monoxide (CO), and that an intermediate in many of the reactions is hydrogen cyanide (HCN). All three compounds are poisonous gases, and it seems paradoxical that they should be forerunners of the earliest biochemistry. They are poisonous, however, only for aerobic forms of life; indeed, for many anaerobes hydrogen sulfide not only is harmless but also is an important metabolite.

It was argued above that oxygen must have been freely available by the time the first eukaryotic cells appeared, probably 1,400 to 1,500 million years ago. Hence the proliferation of cyanobacteria that released the oxygen must have taken place earlier in the Precambrian. How much earlier remains in question. The best available evidence bearing on the issue comes from the study of sedimentary minerals, some of which may have been influenced by the concentration of free oxygen at the time they were deposited. In recent years a number of workers have investigated this possibility, most notably Preston E. Cloud, Jr., of the University of California at Santa Barbara and the U.S. Geological Survey.

One mineral of significance in this argument is uraninite (UO_2), which is found in several deposits that were laid down in Precambrian streambeds. In the presence of oxygen, grains of uraninite are readily oxidized (to U_3O_8) and are thereby dissolved. David E. Grandstaff of Temple University has shown that streambed deposits of the mineral probably could not have accumulated if the concentration of atmospheric oxygen was greater than about 1 percent. Uraninite-bearing deposits of this type are found in sediments older than about two billion years but not in younger strata, suggesting that the transition in oxygen concentration may have come at about that time.

Perhaps the most intriguing mineral evidence for the date of the oxygen transition comes from another kind of iron-rich deposit; the banded iron formation. These deposits include some tens of billions of tons of iron in the form of oxides embedded in a silica-rich matrix; they are the world's chief economic reserves of iron. A major fraction of them was deposited within a comparatively brief period of a few hundred million years beginning somewhat earlier than two billion years ago.

A transition in oxygen concentration could explain this major episode of iron sedimentation through the following hypothetical sequence of events. In a primitive, anoxic ocean, iron existed in the ferrous state (that is, with a valence of $+2$) and in that form was soluble in seawater. With the development of aerobic photosynthesis small concentrations of oxygen began diffusing into the upper portions of the ocean, where it reacted with the dissolved iron. The iron was thereby converted to the ferric form (with a valence of $+3$), and as a result hydrous ferric oxides were precipitated and accumulated with silica to form rusty layers on the ocean floor. As the process continued virtually all the dissolved iron in the ocean basins was precipitated: in a matter of a few hundred million years the world's oceans rusted.

In relation to this hypothesis it is notable that fossil stromatolites first become abundant in sediments deposited about 2,300 million years ago, shortly before the major episode of iron-ore deposition. It is therefore possible that the first widespread appearance of stromatolites might mark the origin and the earliest diversification of oxygen-producing cyanobacteria. Even at that early date the cyanobacteria would probably have released oxygen at a high rate, but for several hundred million years the iron dissolved in the oceans would have served as a buffer for the oxygen concentration of the atmosphere, reacting with the gas and precipitating it as ferric oxides almost as quickly as it was generated. Only when the oceans had been swept free of unoxidized iron and similar materials would the concentration of oxygen in the atmosphere have begun to rise toward modern levels.

Although much remains uncertain, evidence from the fossil record, from modern biochemistry and from geology and mineralogy make possible a tentative outline for the history of Precambrian life. The most primitive forms of life with recognizable affinities to modern organisms were presumably spheroidal prokaryotes, perhaps comparable to modern bacteria of the clostridial type. Initially at least they probably derived their energy from the fermentation of materials that were organic in nature but were of nonbiological origin. These materials were synthesized in the anoxic early atmosphere and were of the type that during the age of chemical evolution had led to the development of the first cells.

The first photosynthetic organisms apparently arose earlier than about three billion years ago. They were anaerobic prokaryotes, the precursors of modern photosynthetic bacteria. Most of them probably lived in matlike communities in shallow water, and they may have been responsible for building the earliest fossil stromatolites known, which are estimated to be about three billion years old.

The rise of aerobic photosynthesis in the mid-Precambrian introduced a change in the global environment that was to influence all subsequent evolution. The resulting increase in oxygen concentration probably led to the extinction of many anaerobic organisms, and others were forced to adopt marginal habitats, such as the lower reaches of bacterial mat communities. Nitrogen-fixing organisms also retreated to anaerobic habitats or developed heterocyst cells. With little competition for those regions having optimum light the cyanobacteria were able to spread rapidly and came to dominate virtually all accessible habitats. With the development of the citric acid cycle and its more efficient extraction of energy from foodstuff, the dominance of the biological community by aerobic organisms was confirmed. When the major episode of deposition of banded iron formations ended some 1,800 million years ago, the trend toward increasing oxygen concentration became irreversible.

By the time eukaryotic cells arose 1,500 to 1,400 million years ago a stable, oxygen-rich atmosphere had long prevailed. Adaptive strategies needed by earlier organisms to cope with fluctuations in the oxygen level were unnecessary for eukaryotes, which were from the start fully aerobic. The diversity of eukaryote cell types present by about a billion years ago suggests that some form of sexual reproduction may have evolved by then. Within the next 400 million years the rapid diversification of

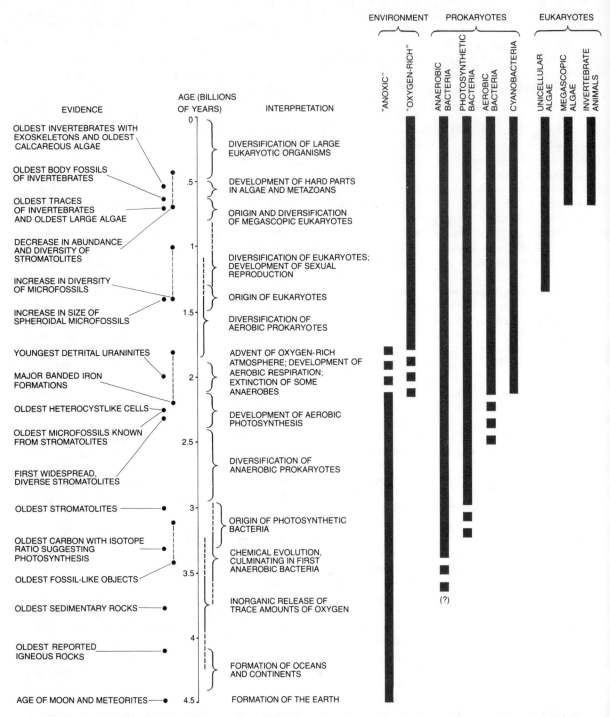

Figure 12 MAJOR EVENTS in Precambrian evolution are presented in chronological sequence based on evidence from the fossil record, from inorganic geology and from comparative studies of the metabolism and biochemistry of modern organisms. Although the conclusions are tentative, it appears that life began more than three billion years ago (when the earth was little more than a billion years old), that the transition to an oxygen-rich atmosphere took place roughly two billion years ago and that eukaryotes appeared by 1.5 billion years ago.

eukaryotic organisms had led to the emergence of multicellular forms of life, some of them recognizable antecedents of modern plants and animals.

In style and in tempo evolution in the Precambrian was distinctly different from that in the later, Phanerozoic era. The Precambrian was an age in which the dominant organisms were microscopic and prokaryotic, and until near the end of the era the rate of evolutionary change was limited by the absence of advanced sexual reproduction. It was an age in which the major benchmarks in the history of life were the result of biochemical and metabolic innovations rather than of morphological changes. Above all, in the Precambrian the influence of life on the environment was at least as important as the influence of the environment on life. Indeed, the metabolism of all the plants and animals that subsequently evolved was made possible by the photosynthetic activities of primitive cyanobacteria some two billion years ago.

The Emergence of Animals

Some 570 million years ago animals diversified at an unprecedented rate. New body plans and ways of living emerged in concert with a new kind of community, one characterized by complex food chains.

. . .

Mark A. S. McMenamin

April, 1987

As recently as the middle of this century the earliest known fossils had all come from geologic formations in a stratigraphic interval known as the Cambrian, which contains rocks formed between about 570 and 505 million years ago. The Cambrian fossils included animals with body plans similar to those of a number of living animals. The fossil record therefore posed a perplexing question: Where were the ancestral forms that had given rise to these plentiful, advanced and diverse early sea animals? The sudden appearance of animal fossils in the lowest Cambrian strata, and the absence of animal fossils in Precambrian strata, made the boundary between the Precambrian and Cambrian periods the cardinal division of the geologic time scale.

In recent years a number of areas throughout the world have indeed yielded animal fossils older than the Cambrian. Still, the fact remains that a tremendous explosion of life took place in the Cambrian period. The start of the Cambrian marks the first appearance of a large number of major groups of animals. Many of these have survived until today. Many others are so unusual that they cannot be assigned to any known phylum. (A phylum is a group of animals that share the same general body plan.) Although most kinds of body plan found in present-day creatures first appeared during the Cambrian period, only a tiny fraction of the animals inhabiting the Cambrian seas gave rise to anything living today. Most of the rest can be viewed, with the benefit of hindsight, as short-lived experiments. The Cambrian period saw a higher percentage of such experimental groups of animals than any other interval in the history of the earth.

The emergence, during the transition from the Precambrian to the Cambrian, of so many kinds of creatures radically changed the nature of the relations among animals. Creatures evolved during the Cambrian to fill ecological niches that had never before been filled. Animals that fed on living matter, rather than scavenging dead organic matter or relying on symbiotic relationships with photosynthesizing algae, became much commoner. Predators emerged. The animals of the Cambrian coexisted in a web of relations not unlike those that connect modern animals. The transition from the Precambrian to the Cambrian thus saw not only the appearance of many modern kinds of animals but also the development of the modern kind of animal community.

The transition from the Precambrian to the Cambrian can be divided into four main stages. The first stage is marked by the appearance of the very first shelly fossils (only one or two shelly species from this stage have been found) and the so-called Ediacaran fauna: flat, soft-bodied creatures named for the Ediacara Hills of South Australia, where many of the first specimens were found. In a very rough approximation, this stage took place in the period between about 700 and 570 million years ago. The second stage, which began about 570 million years ago (with an uncertainty of plus or minus 40 million years), saw the disappearance of the Ediacaran faunas and the first assemblages of shelly faunas of low diversity (in which about five species of shelled animals are found together). This stage lasted for about five to 15 million years.

The third stage, which lasted for about 10 to 20 million years, is characterized by shelly faunas of moderate diversity (more than five but fewer than 15 species of shelled animals found together) and the first appearances of a group of unusual, vase-shaped creatures known as archaeocyathans. The fourth stage, which lasted for some 15 to 30 million years, saw the first shelly faunas of high diversity and the earliest appearance of trilobites. Trilobites are extinct arthropods (animals having jointed appendages) that were made up of three sections (head, thorax and tail) and covered by a shieldlike carapace, or exoskeleton; like modern arthropods, trilobites grew by shedding their carapace, and cast-off carapaces are commonly found as fossils.

The appearance of the Ediacaran fauna, which indicates the first of these four stages, is found in strata above the Precambrian occurrences of glacier-generated marine sediments known as tillites. These tillites record drawn-out episodes of worldwide glaciation. The Ediacaran fauna must therefore have come into being not long after the last major episode of late Precambrian glaciation. Fossils of Ediacaran soft-bodied animals are found with fossil traces (animal tracks and shallow burrows) formed along the surface of what was then the sea floor. There are many fewer fossils at this stage than at later stages, but this can be accounted for partly by the fact that soft-bodied creatures (which decay rapidly) tend not to be preserved as fossils; it does not necessarily indicate that Ediacaran species were not abundant and widespread. A handful of shelled fossils are also found at this stage, including *Cloudina*, which is a tube-shaped fossil from Namibia that has a shell made of calcium carbonate, and *Sinotubulites*, another tubular fossil of calcium carbonate, which is found in Precambrian rocks of southern China.

This interval also yields tube-shaped fossils known as sabelliditids and vendotaenids. Sabelliditids are typically a few centimeters long and from one to several millimeters in diameter. They are fossils of tubular sheaths originally composed of a flexible organic substance that probably housed worm-like, filter-feeding animals. Sabelliditid tubes, along with *Cloudina* and *Sinotubulites*, give evidence of early tube-dwelling animals that had taken up a sessile (stationary), filter-feeding way of life. Vendotaenids are also tubular fossils originally made of a flexible, organic substance, but they are much smaller than sabelliditids (less than a centimeter long and about a tenth of a millimeter wide). Vendotaenids may represent sheaths formed as external secretions by colonies of Precambrian bacteria.

At the beginning of the second stage of the transition period, fossils appear—of animals and pieces of animals—whose hard parts are based on calcium phosphate. The earliest of these phosphatic fossils are a few millimeters or less in length, and they make up a low-diversity grouping of early phosphatic shelly fauna. One example is the tusk-shaped fossil called *Protohertzina*. It has strong microstructural similarities to the sharp spines that a modern phylum of tiny but voracious predators called chaetognaths, or arrow worms, use for grasping. *Protohertzina* itself probably served as a grasping spine of some similar creature, and it is the first fossil that can be identified with any confidence as having been part of a predatory animal. Along with the phosphatic fossils appear some shelly ones consisting of calcium carbonate. The original mineralogy of these early shelly fossils is sometimes difficult to ascertain, because calcium phosphate can replace shell matter made of calcium carbonate with such fidelity that structural features only a few thousandths of a millimeter wide are preserved.

At about the same stratigraphic level there is a tremendous increase in the number and diversity of fossilized tracks and traces on the sea-floor sediment. Deep vertical burrows appear for the first time and there is an abundance of complex trace fossils such as *Phycoides pedum*, a track that records

a series of feeding and burrowing motions made by a bottom-dwelling animal. There are also animal tracks that have chevron-shaped grooved markings, formed as the appendages of a crawling or burrowing arthropod scraped the sediment.

The Ediacaran fauna, with one possible exception, seems to have become extinct at this level, although some paleontologists believe the lack of fossilized corpses of Ediacaran-style softbodied animals may be due to scavenging by burrowing animals, whose number and activity increased markedly at this time.

The third stage of the transition from the Precambrian to the Cambrian is indicated by moderate-diversity shelly faunas. In most areas of the world these appear in strata immediately above those containing the low-diversity faunas. In a region called the Siberian platform, which is in central and western Siberia, the moderate-diversity faunas are accompanied by the earliest archaeocyathans: vase-shaped fossils that had double-walled, porous skeletons made of calcium carbonate. The archaeocyathans bear a slight resemblance to corals and sponges, but in fact they have no close relationship to any living group and are currently placed in their own phylum. Archaeocyathans, in combination with calcareous algae (multicellular algae whose branches and light-gathering organs were reinforced by needles of calcium carbonate), formed the earliest wave-resistant reefs by growing together in mound-shaped accumulations of skeletal material called bioherms.

In the fourth stage of the transition, which is marked by the high-diversity shelly faunas, the geographic range of the archaeocyathans expanded greatly, reaching many areas far from the Siberian platform. In addition, fossils of trilobite carapaces appear for the first time in sediments formed during this stage.

Figure 13 HELICOPLACUS, a creature that became extinct about 510 million years ago, only 20 million years after its first appearance, had what might be called an experimental body plan: its parts were organized in a way that is not found in any living creature. *Helicoplacus* was shaped like a cylindrical spindle and was covered with a spiraling system of armor plates. It emerged during the transition from the Precambrian to the Cambrian period. During that transition more kinds of body plan arose than at any time before or since. Most, like the plan of *Helicoplacus*, proved to be unsuccessful. The specimen shown here is about five centimeters long.

In the 19th century the boundary between the Precambrian and the Cambrian was relatively easy to find, because the areas then under study showed major gaps or unconformities (representing an interval of time in which there was no preservation or deposition of rocks) between Precambrian and Cambrian sedimentary layers. The expanded data base available to 20th-century paleontologists has actually complicated the task of locating the boundary precisely, because many regions are now known where Precambrian and Cambrian formations grade continuously into each other. Determining where the boundary lies in any given rock formation is necessary if paleontologists are to correlate the information gathered at one site with that gathered at another and to determine the relative ages of the formations in question.

Precambrian and Cambrian stratigraphic sections are difficult to correlate from site to site for two reasons. One reason is that there are only a limited number of early fossil species to study. The other is that many of the species that do exist have long stratigraphic ranges (they are found in strata deposited over a broad time interval) and are not particularly useful for splitting stratigraphic sequences into biostratigraphic subdivisions.

Many early fossils are widespread geographically as well as stratigraphically. One such organism is *Anabarites*, a tube-shaped fossil with three distinctive interior ridges. *Anabarites* first appeared during the stage of low-diversity shelly faunas and continued as a component of the high-diversity faunas. It is found in Australia, China, India, Iran, Kazakhstan, Mongolia, western and eastern North America and Siberia. The widespread distribution of some elements of the Ediacaran fauna and the early shelly faunas stems from the positions of the continents in the late Precambrian and early Cambrian. Most of the continents were near the equator, and there is good reason to suppose that in the late Precambrian many of the present-day continents were part of a single supercontinent.

Indeed, Gerard C. Bond and his colleagues at the Lamont-Doherty Geological Observatory have found evidence that continental rifting occurred around the edges of North America during the transition from the Precambrian to the Cambrian. The proximity of the newly separated continents in the early stages of continental drift, and the fact that the continents sat at roughly the same latitude, made it possible for animals to spread widely: there were no

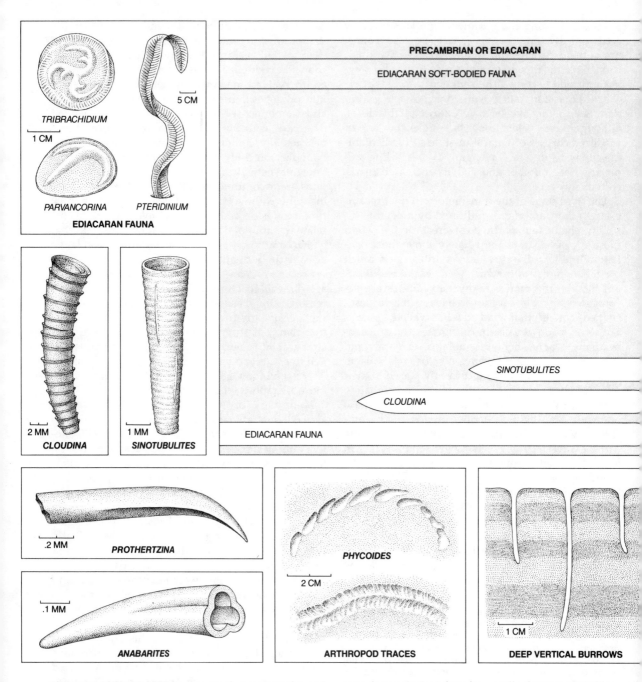

TRIBRACHIDIUM

1 CM

PARVANCORINA

PTERIDINIUM

5 CM

EDIACARAN FAUNA

2 MM

CLOUDINA

1 MM

SINOTUBULITES

PRECAMBRIAN OR EDIACARAN

EDIACARAN SOFT-BODIED FAUNA

SINOTUBULITES

CLOUDINA

EDIACARAN FAUNA

.2 MM   **PROTHERTZINA**

.1 MM

ANABARITES

PHYCOIDES

2 CM

ARTHROPOD TRACES

1 CM

DEEP VERTICAL BURROWS

Figure 14 ANIMAL LIFE became much more complex and diverse during the transition from the Precambrian to the Cambrian period. The Ediacaran fauna lived during the first stage of the transition. A characteristic fossil of the second stage is *Protohertzina*, a relic that was probably a grasping spine of an early predator. Another shelly fossil from the same period is *Anabarites*. In rocks from this period the diversity and number of fossils of the tracks and

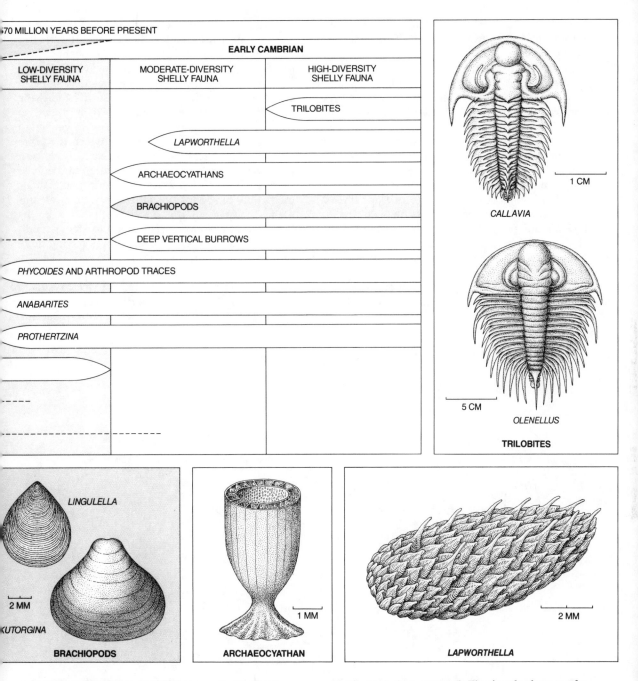

570 MILLION YEARS BEFORE PRESENT

EARLY CAMBRIAN

LOW-DIVERSITY SHELLY FAUNA	MODERATE-DIVERSITY SHELLY FAUNA	HIGH-DIVERSITY SHELLY FAUNA

TRILOBITES

LAPWORTHELLA

ARCHAEOCYATHANS

BRACHIOPODS

DEEP VERTICAL BURROWS

PHYCOIDES AND ARTHROPOD TRACES

ANABARITES

PROTHERTZINA

CALLAVIA

1 CM

5 CM

OLENELLUS

TRILOBITES

LINGULELLA

2 MM

KUTORGINA

BRACHIOPODS

1 MM

ARCHAEOCYATHAN

2 MM

LAPWORTHELLA

shallow burrows left by animals increase markedly. During the next period the brachiopods emerged, as did a strange group of creatures known as archaeocyathans. Toward the middle of this third period *Lapworthella*, an armored animal, first appeared. The fourth phase marks the appearance of the trilobites.

broad oceans or extreme differences in temperature to prevent faunas from migrating from one continental shelf to another.

Provinciality (that is, differences among species in different locales) began to develop throughout the world when the first widespread archaeocyathans and trilobites appeared. The tendency toward provinciality was no doubt intensified by the growth of progressively larger oceans between continents throughout the early Cambrian period. Also, Allison R. Palmer of the Geological Society of America has shown that at this time groups of formations called carbonate belts began to form along the margins of several continents. Carbonate belts are shallow marine areas formed by the accumulation of shells made of calcium carbonate. Because they limit access to the open sea, carbonate belts make it possible for animals living in areas between a belt and the shoreline to evolve in isolation from similar animals living on other continents. The effect is particularly evident in some trilobite groups.

Because most continents in the late Precambrian were near the equator, the climate after the last Precambrian glacial episode was probably quite equable. As the global climate became warmer, food supplies in shallow marine waters began to stabilize at relatively low levels. A decrease in worldwide temperature extremes would have contributed to the stability of marine food resources. In general, a smaller gradient in temperature between the poles and the equator leads to a smaller seasonal overturn of the ocean, causing fewer deep-ocean nutrients to reach shallow marine waters. A stable, nonfluctuating supply of food is crucial for many marine animals, and particularly for those that live in the Tropics and are more accustomed to stable conditions than creatures that live in areas having more pronounced seasonal variations.

As a matter of fact, the limited supply of food available during the Precambrian helps to explain the unusual, flattened bodies of the Ediacaran creatures. The flatness and thinness of the Ediacaran

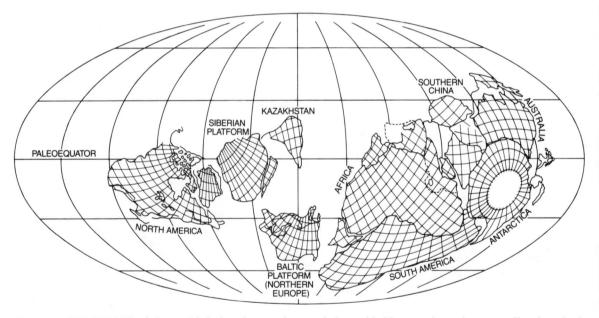

Figure 15 GEOGRAPHY of the world during the transition from the Precambrian to the Cambrian period presented the conditions in which the emerging life of the transition period could flourish. In the late Precambrian much of the earth's landmass was concentrated in a single supercontinent. The breakup of the supercontinent, which probably took place near the beginning of the Cambrian period, provided large regions of new coastline for colonization. Many of the newly formed continents were at or near the equator, and so they offered a warm, equable climate. The increasing distances between the early Cambrian continents encouraged provinciality (differences among separated faunas). Grid lines drawn on the continents mark present-day lines of latitude and longitude.

body plan (the pancake-shaped *Dickinsonia*, for example, had a maximum thickness of no more than about six millimeters although its diameter could exceed one meter!) maximized the ratio of surface area to volume. A high ratio of surface area to volume is ideal for the feeding strategies known as photoautotrophy and chemoautotrophy, which are favored in environments containing small amounts of nutrients.

Photoautotrophy involves a symbiotic relationship with photosynthesizing algae. Within the tissues of the host animal the algae are protected from animals that might feed on them. In return they release nutrients to the host and remove waste products. For such symbiosis to work, a substantial part of the host's body must be exposed to sunlight so that the algae can photosynthesize efficiently. Many reef corals and a few tropical clams harbor symbiotic photosynthetic algae.

The other strategy, chemoautotrophy, is the direct uptake of energy-supplying nutrients from seawater. It too can sometimes involve internal symbiosis, in which the animal harbors chemosynthetic bacteria. The strategy is common, for example, among modern animals living near deep-sea hydrothermal vents. Some, however, seem to absorb dissolved nutrients directly, without the aid of bacteria.

The flattened bodies of the Ediacaran fauna would have made possible efficient uptake of nutrients in seawater or efficient absorption of light by photosynthesizing algae. Recent studies by Pamela Hallock-Muller of the University of South Florida in St. Petersburg have shown that in low-nutrient waters the host-symbiont association is particularly favored, because the host can recycle its waste products directly through the symbiont rather than releasing them into the environment. The Ediacaran fauna may therefore have been well adapted to the marine conditions that prevailed throughout the late Precambrian, when the waters of shallow seas are thought to have been low in nutrients.

Toward the end of the Precambrian and during the transition to the Cambrian the food supply changed, probably in response to changes in the chemistry of the oceans, and other feeding strategies became more important. Heterotrophy—the consumption of other organisms (either animals or plants)—became increasingly important near the end of the Precambrian. Evidence for increased heterotrophy is found, for example, in the fossil record of stromatolites: structures shaped like domes, columns or cones that were built up in layers by seafloor carpets of algae. Stromatolites grew upward toward the sun, successive layers of algal mats rich in organic matter alternating with layers of sedimentary particles that were trapped by the mats. Although each layer was much less than a millimeter thick, the layers gradually accumulated to form large columns or domes; some Precambrian stromatolites were more than 10 meters high.

About 800 million years ago stromatolites underwent a marked decline in diversity, which Stanley M. Awramik of the University of California at Santa Barbara has attributed to the appearance of algae-eating animals: stromatolites are easily overgrazed, and disturbances in the algal mat can cause the stromatolite to stop forming.

Some other possible evidence for increased heterotrophy is found in microfossiliferous cherts, which are sedimentary rocks made up of microcrystalline quartz. When thin sections of these cherts are illuminated and viewed under a microscope, microbial fossils can be seen embedded in the translucent slices of rock. The sheaths of such fossils became more robust between 800 and 700 million years ago, perhaps to provide protection from early grazing animals. At about the same stratigraphic level, fossils of traces made by scavenging and deposit-feeding animals appear. The animals that made these traces were almost certainly the ancestors of the Cambrian shelly organisms.

Perhaps the most astonishing aspect of the Cambrian faunas is that so many radically different types of animals appeared in such a short interval. What gave rise to this sudden diversification? James W. Valentine of Santa Barbara and Douglas Erwin of Michigan State University have proposed that a high order of genetic repatterning was possible because the genome (the full complement of genes) of multicellular animals was less complex in the transition period than it is today. Fewer kinds of mutation would have been fatal because the linkages among various parts of developmental genetic programs were not as intricate in Precambrian and early Cambrian animals as they are in today's fauna.

This genetic flexibility, Valentine and Erwin argue, is one of two reasons so many high-level taxa (such as classes—which are the largest divisions within phyla—and phyla) came into being so suddenly. The other reason is that there were so many niches available that had never before been occupied. The Cambrian could therefore see the appearance of new phyla and classes at a rate that has not been matched since: radically new animals, without competition, could become the founders of new classes or phyla.

The Cambrian diversification, while rapidly establishing new phyla and classes, also initiated the first complex communities of animals linked by food chains. The existence of new types of communities in turn created niches for new types of animals. A key element in the establishment of animal communities is predation, which establishes a hierarchic chain of food-transfer connections among animals. Previous assumptions that predators were not important in Cambrian communities have been overturned by new indications that the impact of predation was great. There are essentially three kinds of evidence: actual fossils of predators, specimens of damaged (and sometimes partially healed) prey and antipredatory adaptations in some animals.

Protohertzina, the fossil that resembles a modern arrow worm's grasping spine, is a relic of one early Cambrian predator. Another predator is *Anomalocaris*, which has recently been reconstructed from nearly complete specimens by Derek Briggs of the University of Bristol and Harry Whittington of the University of Cambridge. *Anomalocaris* is gigantic by Cambrian standards (about 45 centimeters long) and resembles no living animal. It had a body shaped like a flattened teardrop, both sides of which were flanked by swimming fins. At its broad head were a pair of jointed appendages for drawing prey into its hideous mouth, which was shaped like a pineapple ring lined with teeth. *Anomalocaris* is probably an example of a short-lived, experimental phylum. Briggs and Whittington suggest that *Anomalocaris* is largely responsible for one of the other major signs of predation in the Cambrian: fossils of wounded trilobites.

There are numerous fossilized trilobite specimens that have had bites taken out of the carapace. In most cases the wounds are partially healed, indicating that the carapace was still attached to the trilobite when it was damaged, rather than having been damaged by a scavenger after being shed by the trilobite. There are other examples of predatory damage, such as small shelly fossils with boreholes in them. The holes resemble those made by certain modern predators that drill through shells to eat the soft meat inside.

The third indication that predation was important in the early Cambrian is the number of features of early Cambrian animals that could have had antipredatory functions. For example, the shells and exoskeletons that appear for the first time in a number of the new phyla probably served as a defense against predators. The deep vertical burrows that proliferate at the time of the low- and moderate-diversity shelly faunas could have given protection from predators unable to move through sediment. Several trilobite species evolved long spines, which may have made them more difficult for such contemporary predators as *Anomalocaris* to attack.

In addition, Stefan Bengtson of the Insitute of Paleontology in Uppsala, Ed Landing of the New

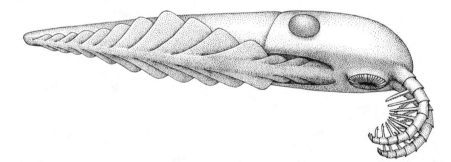

Figure 16 PREDATOR from the early and middle Cambrian period may have fed largely on trilobites. This creature, *Anomalocaris*, was much larger than most animals of its day (it grew to be about 45 centimeters long). It had gripping appendages to carry food to its mouth; it swam by generating a wavelike motion along the finlike membranes on its underside.

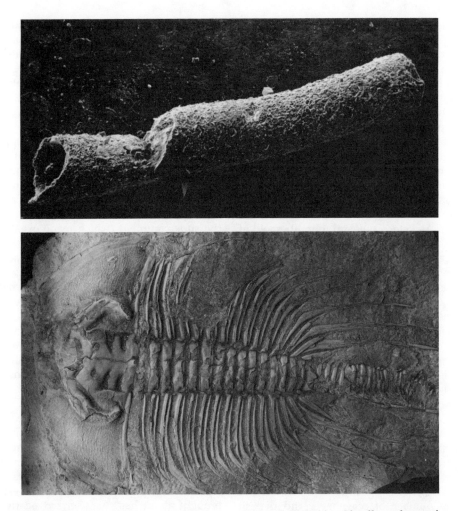

Figure 17 WOUNDED ANIMALS provide striking evidence of predation during the transition period. The image at the top shows a damaged and healed shell of *Hyolithellus* enlarged some 40 diameters. The photograph at the bottom shows a wounded and healed carapace (exoskeleton) of the trilobite *Olenellus robsonensis* seen at three-fourths actual size. That the wounds have healed is important, because it demonstrates that the damage was done during the life of the animal and was not inflicted later on a corpse or an empty shell.

York State Geological Survey and Simon Conway Morris of Cambridge have shown that many small shelly fossils are actually sclerites: disarticulated parts of a spiny, maillike coat of armor that probably protected the upper surface of slow-crawling animals. Such animals may have resembled small marine porcupines or hedgehogs. Another adaptation is found in the class of brachiopods (creatures with a two-valved shell roughly resembling that of clams) referred to as mickwitziids. These animals had numerous pores, called punctae, running through their shell walls. The punctae may have delivered chemical deterrents to the outer surface of the shell, discouraging predators and parasites.

Predators, then, were an important part of the marine environment during the Cambrian. Exoskeletons, which may originally have evolved as a means of protection, were also key factors in the evolution of some strikingly new body plans. For example, without a bivalved shell brachiopods would not be able to produce the internal currents that make their filter-feeding mechanism efficient.

Even though the mechanics of the establishment of Cambrian communities are becoming clearer, one central question remains unanswered: Why did the Cambrian revolution occur when it did and not tens or even hundreds of millions of years earlier? This question is particularly puzzling because scavenging and grazing animals (known from their trace fossils) predate the Cambrian boundary by as much as 200 million years.

One answer may have to do with the chemistry of the oceans. The concentrations in seawater of phosphate, as well as of isotopes of sulfur and strontium, underwent dramatic (if poorly understood) fluctuations during the transition from the Precambrian to the Cambrian. As Peter Cook and John Shergold of the Bureau of Mineral Resources, Geology and Geophysics in Canberra suggest, the large phosphate deposits found in Precambrian-Cambrian sediments in many parts of the world may represent an episode of global phosphogenesis, when increases in the availability of phosphate and other nutrients made it easier for animals to form phosphatic skeletons.

A stumbling block to this hypothesis is that calcareous shells seem to be at least as prevalent as phosphatic shells at the boundary between the Precambrian and the Cambrian. It may be more accurate, then, to see the episode of phosphate deposition as part of a larger event in which the oceans saw a sudden increase in the amount of available nutrients. An environment rich in nutrients would no longer have favored the bioenergetics of the host-symbiont relationship and so it would have increased the number of scavengers and deposit feeders, leading in turn to an increase in the number of predators. As such heterotrophic organisms reached what might be called a critical biomass, an "ecological chain reaction" (in the words of Martin Brasier of the University of Hull) could be initiated: newly evolved animals might create niches that could be filled by other, newer animals, eventually resulting in complex communities in which there were abundant shell-bearing animals.

The Ediacaran fauna (and the other Precambrian animals) had originated in a world characterized by a massive supercontinent, waning glaciation and relatively low supplies of marine nutrients. The remarkably numerous and diverse Cambrian animals appeared in a world marked by the breakup of the supercontinent (which made extensive tropical shorelines available) and abundant marine food supplies. It is not yet known with certainty whether the Cambrian animals appeared because of such changes in the global environment, because of a series of fortuitous changes in the genetic programming of animals or because of some combination of these causes and other, unrelated factors.

Whatever the reason for their origin, the earliest Cambrian innovations (such as shelly organisms, predators and deep burrowers) went on to colonize the world rapidly. The combination of environmental influences (such as nutrient-rich waters) and biotic changes (such as the emergence of predators) caused a major change in the nature of animal communities. Modern animals, including human beings, are the direct descendants of the animals that first appeared during the Cambrian explosion, and the style of ecological interaction these early animals brought about has characterized nearly all animal communities of the past 570 million years.

SECTION

CHALLENGES IN THE SEA

. . .

Introduction

...

Life began in the water, and so it is fitting that it is with the sea that we should begin our look at species on or (we might have thought) over the edge. Since chemoautotrophy was the lifestyle of the first of the earth's organisms, it is especially appropriate that we start with one of the biggest biological surprises of the decade: the discovery of life in the boiling and toxic waters near undersea volcanic vents. Although finding life at such depths has always been surprising [see "Animals of the Deep-Sea Floor," by John D. Isaacs and Richard A. Schwartzlose; SCIENTIFIC AMERICAN, October, 1975; and "Microbal Life in the Deep Sea," by Holger W. Jannasch and Carl O. Wirsen; SCIENTIFIC AMERICAN, June, 1977], the organisms found earlier were scavengers, living on organic matter that sank from near the surface. As Chapter 3, "Symbiosis in the Deep Sea," points out, the vent fauna live on energy produced at the bottom. Their energy source is hydrogen sulfide (H_2S), which they "burn" with oxygen to produce water and various sulfates. Almost any reaction that produces water liberates vast amounts of energy; the space shuttle, in one sense the world's largest steam engine, combines oxygen and hydrogen to produce water and enormous heat. The energy liberated by those oxidation reactions is originally stored in the high-energy electrons of the unoxidized hydrogens, in compounds referred to by chemists and biologists as "reduced."

Although the energy is there for the harvesting, hydrogen sulfide, like cyanide, poisons oxygen-dependent respiration by combining with and deactivating one of the enzymes involved in charging the mitochondrial battery. Yet life flourishes by the vents, at temperatures that would cook most creatures and in water acid enough to dissolve most others. [For more on the chemistry of vent waters see "Hot Springs on the Ocean Floor," by John M. Edmond and Karen Von Damm; SCIENTIFIC AMERICAN, April, 1983]. Interestingly enough, each of the species studied to date avoids the hydrogen sulfide poisoning in a different way, indicating that natural selection has followed up different "preadaptations" to produce organisms able to survive in this inhospitable habitat. The intimate symbiosis between bacteria and vent worms represents the most highly evolved strategy for beating this particular set of odds.

We have talked loosely of challenging niches and demanding habitats without making a formal distinction. The world has been divided up by natural selection into habitats and niches, categories defined as much by who occupies them as what they are. A habitat is *where* an animal makes its living, while its niche is *how*—its occupation. To support life a habitat must allow access to the energy and materials essential for organisms, but how the species in a habitat evolve to use and in some sense "share" these crucial resources—their niches—can be quite different in two identical but separate habitats. Often the constellations of species we see in one place today are there because of chance events that led some to colonize and adapt to the local situation before others, better adapted initially, arrived on the scene. Other times a trivial feature of one species preadapted it to a climatic change or habitat disruption that wiped out its competition. Indeed, there are some species that thrive on the sorts of variability and environmental chaos that regularly drive other creatures to local extinction.

In Chapter 4, "Antarctic Fishes," the habitat is the subzero waters off that coldest of continents. That is also a remarkably energy-poor environment: below the ice, the light level available to autotrophs is about 1 percent of normal in the summer, and during the winter there are months on end of total darkness. The problem facing the fish is to conserve energy and avoid freezing. The latter task is especially serious since even small ice crystals can destroy delicate cell membranes. The solution for the fish lies in exploiting the polar nature of water, using a novel antifreeze that interferes with icing. The origin of this preadaptation is unknown—that is, we have no idea what the evolutionary precursor of the antifreeze used to do—but since 90 percent of the Antarctic species belong to one suborder, all probably arose from one or at most a few closely related species that possessed this preadaptation when the continent began to cool. The habitat has since been subdivided into a variety of niches based primarily on depth.

In Chapter 5, "Intertidal Fishes," we see an equally challenging habitat, but one as variable as the Antarctic environment is stable and predictable. With each cycle of the tides the denizens of the shoreline risk being battered by the surf, exposed to the air and predators, being trapped in a pool of extremely high or low salinity (either equally fatal to most fish) at threateningly high or low temperatures or being trapped in a pool so small that the creatures' own metabolism fills it with otherwise fatal levels of carbon dioxide. Yet the intertidal zone is filled with species specialized in different ways to survive where life seems impossible—a habitat where the first terrestrial organisms may have gotten their unexpected initiation into life on land.

Symbiosis in the Deep Sea

The remarkable density of life at deep-sea hydrothermal vents is explained by the mutually beneficial symbiosis of invertebrate animals and sulfide-oxidizing bacteria that colonize their cells.

. . .

James J. Childress, Horst Felbeck and George N. Somero

May, 1987

Biologists categorize many of the world's environments as deserts: regions where the limited availability of some key factor, such as water, sunlight or an essential nutrient, places sharp constraints on the existence of living things. Until recently the deep sea was considered to be such a desert, where the low abundance of organisms stems from the extreme limitation of the food supply.

Yet there is one habitat in the deep sea where the density of life equals, if it does not surpass, what is found in any other marine ecosystem. It is the system of hydrothermal vents, or deep-sea hot springs, situated at sea-floor spreading centers. The vents, discovered only 10 years ago, are found along ridges at the bottom of the ocean where the earth's crustal plates are spreading apart.

It was at a spreading ridge in the Pacific Ocean, 320 kilometers northeast of the Galápagos Islands, that geologists on the research submarine *Alvin* in 1977 discovered an oasis of densely packed animal life 2,600 meters below the surface. Here were species previously unknown to science, living in total darkness in densities enormously higher than had ever been thought possible in the deep sea. Giant tube worms as much as one meter long, large white clams 30 centimeters in length and clusters of mussels formed thick aggregations around the hydrothermal vents. There were smaller but nonetheless significant numbers of shrimps, crabs and fishes.

Such biological density was completely unexpected, and also very puzzling. Any ecosystem depends on the presence of primary producers: autotrophic (self-supporting) organisms that synthesize their own reduced carbon compounds, such as carbohydrates, from carbon dioxide. Green plants that convert carbon dioxide into carbohydrates in the presence of sunlight are called photoautotrophs, and they are the primary producers in most marine and terrestrial ecosystems.

Yet clearly photosynthesis is impossible at the depth of the vents. Not enough light penetrates beyond the upper 200 or 300 meters of the ocean's surface to support photosynthesis. Below the sunlit surface the density of life falls off rapidly because there is less food. The organisms that live in the

deep sea depend on organic matter that drifts down from the sunlit "euphotic" zone to the ocean floor. Most organic matter is synthesized, consumed and recycled in the euphotic zone, and only a small fraction sinks to the deep sea. Yet at the vents life was blooming 2,600 meters below the surface of the ocean. What unique feature of the hydrothermal-vent system could explain the abundance of life at these depths?

I t seemed possible that the density of life at these deep-sea hot springs might be explained on the basis of temperature. In contrast to most of the deep sea, which is very cold (from two to four degrees Celsius), the vent waters have average temperatures ranging from 10 to 20 degrees. Observations made over the past six years have shown, however, that the warm temperatures do not account for the unique fauna of the vent ecosystem.

Instead there is a better explanation for these teeming oases of life at the bottom of the ocean. Water samples taken from the vent sites and analyzed by John M. Edmond of the Massachusetts Institute of Technology and his colleagues indicated that these underwater hot springs, like many hot springs on land, are rich in hydrogen sulfide. This energy-rich but highly toxic compound is found at high levels in the water that flows from cracks in the ocean floor [see "Hot Springs on the Ocean Floor," by John M. Edmond and Karen Von Damm; SCIENTIFIC AMERICAN, April, 1983].

For years biologists had known that sulfide-rich habitats, such as terrestrial hot springs, support large numbers of free-living bacteria. Like green plants, they are autotrophs, but they get energy for carbon fixation—the conversion of carbon dioxide into organic molecules that serve as nutrients—not from the sun but from the oxidation of hydrogen sulfide.

Figure 18 **HYDROTHERMAL-VENT COMMUNITIES** consist of an array of unusual invertebrate species. Dense clusters of tube worms as much as one meter long are anchored to the basaltic rock along vent openings. Living among them are vent mussels (*yellow*) and vent crabs. (The crab near the center of the photograph may be attempting to feed on a tube worm.) Smaller invertebrates can be seen clinging to the worms and to the shells of the mussels. The photograph was made from the research submarine *Alvin* by R. R. Hessler of the Scripps Institution of Oceanography on an expedition to the Galápagos Rift vent site in the East Pacific Ocean.

Holger W. Jannasch of the Woods Hole Oceanographic Institution and David Karl of the University of Hawaii at Manoa isolated bacteria from the vent waters and did a series of experiments demonstrating that some species of vent bacteria are indeed autotrophic and depend on hydrogen sulfide and other reduced forms of sulfur for their metabolic activities. These sulfide-oxidizing bacteria might form the base of the vent food chain by providing food for the animal species.

Although they are very distantly related, green plants and sulfur bacteria are functionally similar in an important way: both are primary producers. Whereas green plants are photoautotrophs, sulfur-oxidizing bacteria capable of using an inorganic energy source to drive carbon dioxide fixation are called chemolithoautotrophs: they are literally consumers of inorganic chemicals that can exist autonomously—without an external source of reduced carbon compounds. Both fix carbon dioxide through a series of biochemical reactions collectively known as the Calvin cycle. The Calvin cycle functions in much the same way for bacteria as it does for green plants; the end product for both is reduced carbon.

I t appeared, then, that the sulfur bacteria were in effect serving as the "green plants" of the vents and that their "sunlight" was hydrogen sulfide and other reduced carbon compounds. A major complication of this simple model was soon recognized: one of the dominant vent animals, the tube worm *Riftia pachyptila*, seemed to lack any means of harvesting the abundant crop of sulfur bacteria proliferating in the waters all around it.

Riftia is a strikingly unusual creature by conventional anatomical standards. It is essentially a closed sac, without a mouth or a digestive system and with no other means of ingesting particulate food. At its anterior tip there is a bright red branchial (gill-like) plume where oxygen, carbon dioxide and hydrogen sulfide are exchanged with the ambient seawater. Below the plume there is a ring of muscle, the vestimentum, that anchors the worm in its white tube. Most of the rest of the animal consists of a thin-walled sac that contains the worm's internal organs. The largest of them is the trophosome, which occupies most of the body cavity. As its name ("the feeding body") suggests, the trophosome contributes significantly to the worm's nutrition—but it lacks a channel through which particulate materials from the outside world can enter the worm. The big

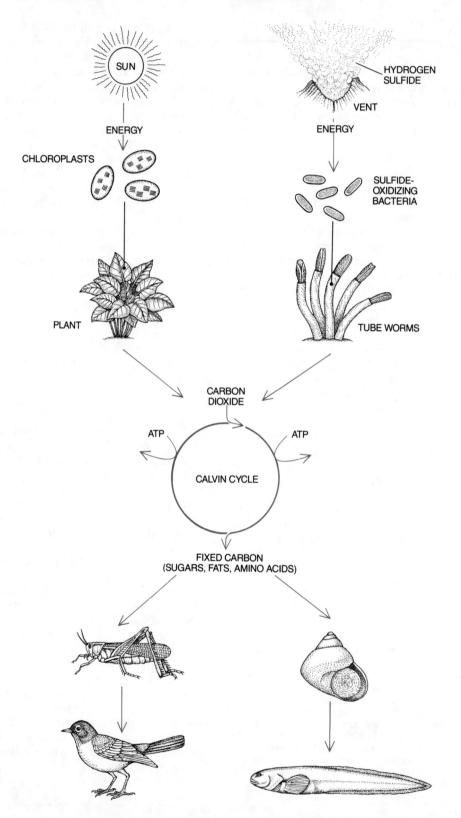

PHOTOSYNTHESIS

CHEMOSYNTHESIS

SUN

HYDROGEN SULFIDE

VENT

ENERGY

ENERGY

CHLOROPLASTS

SULFIDE-OXIDIZING BACTERIA

PLANT

TUBE WORMS

CARBON DIOXIDE

ATP

ATP

CALVIN CYCLE

FIXED CARBON
(SUGARS, FATS, AMINO ACIDS)

question was how, in view of its unusual anatomy, *Riftia* manages to obtain the nutrients it needs for survival.

Microscopic studies done by Colleen M. Cavanaugh of Harvard University and her colleagues and parallel biochemical studies done by us provided the first clues. Examination of the trophosome of *Riftia* revealed that it is colonized by vast numbers of sulfur-oxidizing bacteria. We recognized that the bacteria and *Riftia* had established what is known as an endosymbiotic relation.

Symbiosis is the co-occurrence of two distinct species in which the life of one species is closely interwoven with the life of the other. Symbioses vary from relations that are beneficial to one partner and harmful to the other (parasitism) to relations from which both the partners benefit (mutualism). When one species, known as the symbiont, lives within the body of the other, known as the host, the relation is called endosymbiotic.

The *Riftia*-bacteria endosymbiosis is mutualistic. The tube worm receives reduced carbon molecules from the bacteria and in return provides the bacteria with the raw materials needed to fuel its chemolithoautotrophic metabolism: carbon dioxide, oxygen and hydrogen sulfide. These essential chemicals are absorbed at the plume and transported to the bacteria in the trophosome by the host's circulatory system. The worm's trophosome can be thought of as an internal factory, where the bacteria are line workers producing reduced carbon compounds and passing them to the animal host to serve as its food.

The ability of *Riftia* to absorb sulfide from vent water and transport it to the bacteria in its trophosome without either poisoning itself or degrading the sulfide presented a major puzzle. Hydrogen sulfide is a highly toxic compound, comparable to cyanide in its ability to block respiration, the process

Figure 19 PHOTOSYNTHESIS and chemosynthesis are compared. The energy sources differ, but the conversion process and end products are the same. In photosynthesis light is absorbed by the chloroplasts of plants and drives carbon fixation by the Calvin cycle, a process yielding sugars, fats and amino acids that enter the food chain, passing from herbivores to carnivores. In chemosynthesis energy is provided by hydrogen sulfide issuing from vents in the ocean floor. It is taken up by free-living bacteria and also absorbed by vent animals such as the tube worm, which transport it to endosymbiotic bacteria. In the bacteria it is oxidized, providing energy for the Calvin cycle. The end products enter the food chain, passing directly from lower-order carnivores to higher-order ones.

whereby the animal uses oxygen. In most animals sulfide inhibits respiration in two ways: by blocking oxygen's binding sites on its major carrier, the hemoglobin molecule, and by poisoning an important respiratory enzyme, cytochrome *c* oxidase. Studies of *Riftia* showed, however, that sulfide has no effect on oxygen binding and that the worm's respiration rate is substantial even in the presence of sulfide concentrations lethal to most animals.

We wanted to know how aerobic respiration in *Riftia* is possible in the presence of high sulfide concentrations. Clearly *Riftia* had three obstacles to overcome. First, it needed to evolve a special transport system to extract sulfide from vent water. Second, it needed to transport sulfide in its blood without allowing the sulfide to compete with oxygen for binding sites on the hemoglobin molecule or to react with oxygen. (In the presence of oxygen, sulfide is highly unstable, decomposing rapidly to oxidized forms such as thiosulfate and elemental sulfur.) Third, it needed a mechanism to prevent sulfide from diffusing into its cells and poisoning respiration.

We collaborated with Mark A. Powell (who is now at the University of California at Davis) and Steven C. Hand (now at the University of Colorado at Boulder) to isolate cytochrome *c* oxidase from plume cells and examine its behavior in the presence of sulfide. Cytochrome *c* oxidase is responsible for the final step in the chain of metabolic reactions known as oxidative phosphorylation, the most important process by which adenosine triphosphate (ATP), the major energy currency of the cell, is synthesized in aerobic (oxygen-using) organisms. In most animals minute concentrations of sulfide are enough to inhibit cytochrome *c* oxidase. We first hypothesized that *Riftia* might have evolved a sulfide-insensitive form of the enzyme. Experiments showed this is not the case: highly purified cytochrome *c* oxidase from *Riftia* is just as sensitive to sulfide poisoning as cytochrome *c* oxidase from other animals.

We noted in experiments with the enzyme that its sensitivity to sulfide depended on the extent to which we isolated it from other proteins of the plume. As we went through successive purification steps we observed a substantial decrease in the bright red color of our preparation, and with it an increase in the sensitivity of cytochrome *c* oxidase to sulfide poisoning. The color change suggested that something in the blood was protecting cytochrome *c* oxidase from the toxic effects of sulfide. We proved

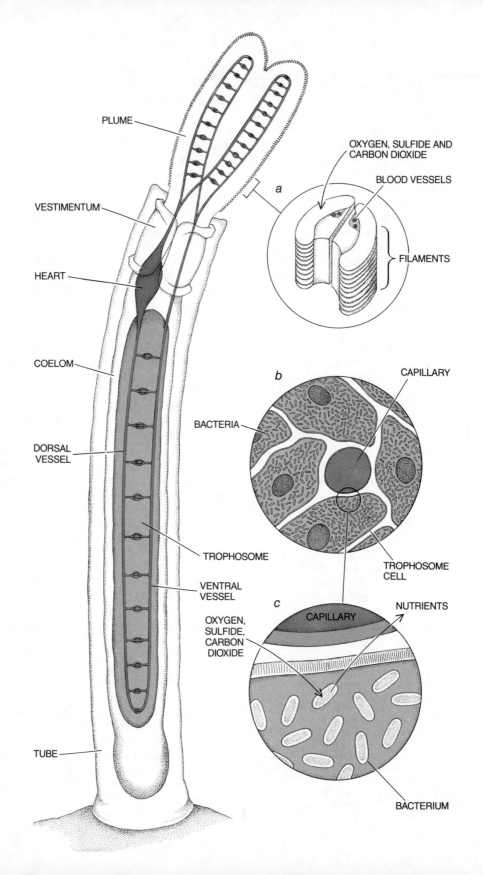

PLUME

VESTIMENTUM

HEART

COELOM

DORSAL
VESSEL

TROPHOSOME

VENTRAL
VESSEL

TUBE

a

OXYGEN, SULFIDE AND
CARBON DIOXIDE

BLOOD VESSELS

FILAMENTS

b

CAPILLARY

BACTERIA

TROPHOSOME
CELL

c

CAPILLARY

NUTRIENTS

OXYGEN,
SULFIDE,
CARBON
DIOXIDE

BACTERIUM

this by adding a minute amount of whole blood to a poisoned enzyme system. Almost immediately the cytochrome *c* oxidase activity returned to a normal, uninhibited level. The observation showed that a blood component—perhaps hemoglobin—was binding sulfide more strongly than the sulfide-sensitive cytochrome *c* oxidase system was, thereby preventing poisoning of respiration. This discovery and simultaneous studies of the worm's hemoglobin done during a 1982 expedition to the East Pacific Rise vent site indicated that hemoglobin is indeed a key player in the worm's internal transport system.

Riftia has a rich blood supply: the deep red of the branchial plume is due to the presence of a large volume of hemoglobin-rich blood, which accounts for more than 30 percent of the worm's total volume. The total concentration of hemoglobin per liter—approximately half what it is in human blood—is extremely high for an invertebrate. Moreover, *Riftia* hemoglobin is very different from human and other vertebrate hemoglobins. It is large, with a molecular weight of as much as two million daltons (human hemoglobin has a molecular weight of 64,000 daltons). Rather than being contained within red blood cells, it circulates freely in the serum. It also has an unusually high affinity for oxygen as well as an unusually high carrying capacity for oxygen. *Riftia* hemoglobin is thus well adapted for extracting oxygen from the vent waters and transporting it to the cells of the tube worm and to its symbiont.

There is an even more important and striking difference between the hemoglobin of *Riftia* and other hemoglobins: the tube-worm molecule can bind both oxygen and hydrogen sulfide simultaneously. This discovery, made by Alissa J. Arp (who is now at San Francisco State University) and us, suggested that the site where sulfide binds to the molecule is different from the site where oxygen binds. Charles R. Fisher, Jr., of the University of

California at Santa Barbara showed that the *Riftia* hemoglobin molecule stabilizes mixtures of oxygen and sulfide, preventing the spontaneous oxidation of sulfide. We therefore concluded that hemoglobin plays a dual role in the tube worm: it prevents sulfide from poisoning respiration and it also protects sulfide by allowing it to be transported to the trophosome without being oxidized.

After it is unloaded in the cells of the trophosome, the sulfide is oxidized. The oxidation takes place in the bacterial symbionts, as was demonstrated by a special staining procedure that signals the presence of hydrogen sulfide. Further investigation by our group showed that the sulfide oxidation drives the synthesis of ATP and the fixation of carbon dioxide. Moreover, we showed that the activity of the Calvin cycle is approximately the same in the trophosome bacteria as it is in the leaves of a green plant.

Studies of other vent animals reveal that *Riftia pachyptila* is not the only hydrothermal-vent species that has evolved a symbiotic relation with sulfur bacteria. The large white clam, *Calyptogena magnifica,* and the mussel, *Bathymodiolus thermophilus,* also depend on chemosynthetic endosymbionts for food. But these species have evolved quite different approaches to the same problem.

In *Calyptogena* the bacteria are not in an internal organ but in the gills, where they can readily obtain oxygen and carbon dioxide from the respiratory water flow. The basic metabolic plan is the same, however: the bacteria oxidize sulfide and supply the clam with fixed carbon compounds. Like other invertebrates harboring sulfur bacteria as endosymbionts, *Calyptogena* has a greatly reduced ability to feed on and digest particulate foods.

Still, we were puzzled at first by the data from water and blood samples. The clams must be concentrating sulfide in their blood, because the sulfide level there is orders of magnitude higher than the concentration in the ambient water. Because of the way the clams are oriented in the vent waters, however, the sulfide concentration in the water bathing their gills is low. From what source, then, do these giant clams obtain sulfide to feed their endosymbionts? The path turns out to be indirect. Apparently the clams absorb sulfide through their large, elongated feet, which extend into the hydrothermal vents, where the concentrations of sulfide are highest. Once it is absorbed through the clam's feet, sulfide is transported by the blood to the bacteria in the gills.

Figure 20 TUBE WORM *Riftia pachyptila* **is anchored inside its protective outer tube by a ring of muscle, the vestimentum. At its anterior end is a respiratory plume. Oxygen, sulfide and carbon dioxide absorbed through the plume filaments (***a***) are transported in the blood (***red***) to the cells of the trophosome (***blue***). The trophosome, site of chemosynthesis, represents one-sixth of the mass of the animal and fills up much of the coelom, or body cavity. Dense colonies of endosymbiotic bacteria live within the trophosome cells (***b***). Oxygen, sulfide and carbon dioxide pass from the capillaries of the worm to the bacteria (***c***). Nutrients pass from the bacteria to the capillaries for distribution throughout the animal.**

Figure 21 LARGE VENT CLAMS, *Calyptogena magnifica,* cluster along cracks in the sea floor where sulfide-rich water is vented at a site on the East Pacific Rise. The clams' feet penetrate the cracks, where the sulfide concentration is highest. Cobweblike growths on the rocks are colonies of free-living sulfur-oxidizing bacteria. The painting is based on a photograph made by Kenneth L. Smith, Jr.

Transport of sulfide in the blood of *Calyptogena* is a very different process from the one described for *Riftia.* The clam's hemoglobin (which is incorporated within the animal's red blood cells) is irreversibly poisoned by sulfide, and so it cannot serve as a sulfide-transport protein. *Calyptogena* has overcome the problem of sulfide poisoning by evolving a special sulfide-transporting protein. It is an extremely large protein and it circulates in the serum rather than in the red cells. The protein has a dual func-

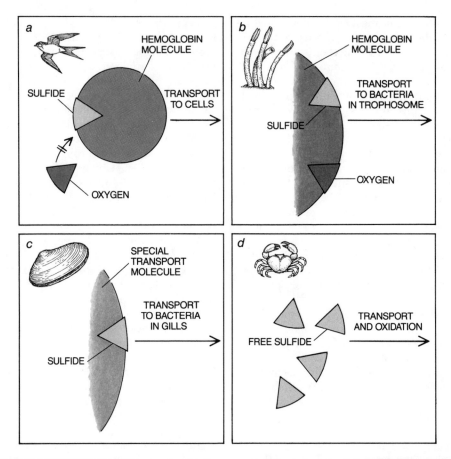

Figure 22 SULFIDE IS HIGHLY TOXIC to most animals (*a*). It poisons respiration at two levels: in the blood, where it binds to hemoglobin, and in cells, where it inhibits the respiratory enzyme cytochrome *c* oxidase (not shown). Animals associated with sulfide-rich hydrothermal vents have evolved different strategies to avoid sulfide poisoning. The tube worm *Riftia pachyptila* (*b*) has a separate binding site on its hemoglobin molecule for sulfide and so can transport oxygen and sulfide simultaneously in its bloodstream. The vent clam *Calyptogena magnifica* (*c*) has a special transport protein to carry sulfide to bacteria in its gills. The vent crab *Bythograea thermydron* (*d*) lacks endosymbiotic bacteria; it is able to detoxify sulfide by oxidizing it to nontoxic thiosulfate in its liverlike hepatopancreas.

tion: it protects the animal's hemoglobin and cytochrome *c* oxidase from sulfide poisoning and also protects sulfide against oxidation on the way to the gills. The binding of sulfide to the protein is reversible: it is off-loaded to bacteria in the gill (by mechanisms not yet understood), where it is oxidized to provide energy for the Calvin cycle.

Much less is known about symbiosis in *Bathymodiolus thermophilus*. Like the giant clam, the mussel has bacterial symbionts in its gills, but the pathways of sulfide transport and metabolism are not yet understood. Kenneth L. Smith, Jr., of the Scripps Institution of Oceanography has, however, carried out

an interesting series of experiments demonstrating that the relation between these deep-sea mussels and the vent water is obligatory. When he moved mussels away from the immediate vicinity of a vent to a more peripheral region, they showed clear signs of starvation. Such starvation is a naturally occurring process at hydrothermal-vent sites. Individual vent sites are active for only a few decades at most. Large numbers of dead mussels and clams litter sites where water flow has ceased, indicating that for these animals survival is not possible without a supply of sulfide.

Growth rate serves as another indication that

symbiosis is effective in meeting the nutritional requirements of the animal hosts. The rapid turnover of life at these vent sites is reflected in the accelerated growth and the quick attainment of reproductive age that are observed in these animals. Work done by Richard A. Lutz of Rutgers University shows that the clams and mussels of the vents grow as fast as the fastest-growing bivalves found in shallow waters.

The tube worm, the clam and the mussel, the largest and most dominant numerically of the vent species, owe their ecological success to their symbiosis with sulfur-metabolizing bacteria. Many of the smaller and less conspicuous vent animals lack symbionts. They obtain their nutrients either by filtering particulate food, such as bacteria, from the water or by feeding on the animals that do contain symbionts. We have observed vent crabs, for example, in the act of feeding on the respiratory plume of *Riftia*.

The symbiont-free animals are interesting objects to study in their own right. Because sulfide is readily absorbed across the surface of an invertebrate's body, the symbiont-free animals, like the symbiont-containing species, have had to evolve mechanisms to prevent sulfide poisoning. With the collaboration of Russell Vetter of Scripps and Mark Wells of Santa Barbara, we examined these mechanisms in the vent crab.

The animal's heart rate and the beating of its scaphognathites (appendages that drive the flow of respiratory water) did not change when we raised sulfide concentrations well above normal levels in ambient seawater, and the level of sulfide in the crab's blood increased only minimally. Because we could not isolate from the crab's blood a sulfide-binding protein comparable to the tube worm's hemoglobin, we assumed this species must have evolved some different strategy for detoxifying sulfide. We found that the crab can detoxify the sulfide

it absorbs by oxidizing sulfide to thiosulfate, a much less toxic form of sulfur. The process is carried out in the crab's hepatopancreas, a tissue similar in function to the vertebrate liver.

We wondered whether, in the symbiont-containing animals, each host was colonized by only a single type of bacterial symbiont. Recent studies of the sequences of a particular ribonucleic acid (the 16S RNA) of bacterial ribosomes, done by Daniel Distel and our group in collaboration with Norman R. Pace of Indiana University, indicate that the relation between a sulfur-metabolizing bacterium and its animal host is species-specific, that is, each host species harbors a unique strain of bacteria. In spite of the fact that many types of sulfur bacteria have entered into this kind of symbiosis, no one bacterium seems to have adapted itself to more than one species of host. Presumably endosymbiosis with sulfur bacteria originated independently and repeatedly in diverse animal groups.

Our discoveries at deep-sea vents have stimulated surveys of a wide range of sulfide-rich environments, including mangrove swamps, petroleum seeps, sewage outfall zones and marshes. The surveys have revealed that the kinds of sulfur-based symbioses first discovered in the deep sea are widespread, and that various other energy-rich inorganic molecules, such as methane, can be exploited in analogous symbioses between other invertebrates and bacteria. These diverse symbioses have been described in disparate animal groups, including some smaller tube-worm relatives of *Riftia*, clams of a number of different families, various mussels and several groups of small marine worms.

As our studies progress we expect to find further instances in which animals rely on chemolithoautotrophic symbionts, thereby gaining the ability to tolerate and exploit habitats the animals could not colonize successfully without bacterial partners.

Antarctic Fishes

Most fish species perished when the Antarctic Ocean turned cold and icy, but fishes of one suborder, Notothenioidei, met the challenge. They survive by making biological antifreezes and conserving energy.

. . .

Joseph T. Eastman and Arthur L. DeVries

November, 1986

In February of 1899 the British ship *Southern Cross* put 10 men ashore at Cape Adare in Antarctica, thus beginning the first expedition to endure a year on the world's southern-most continent. Many zoologists credit the expedition, which ushered in the "heroic" era of Antarctic exploration, with a discovery that has intrigued them for almost a century: the coldest marine habitat in the world is actually alive with fishes. The team's zoologist, Nicholai Hanson, did not survive the year on the icy land, but before his death he collected examples of previously unknown fish species.

Almost a century later we and other investigators are still trying to understand the adaptations that enable fish to survive in a region once thought to be virtually uninhabitable. Of particular interest are the evolutionary adaptations of the suborder Notothenioidei, a group of teleosts, or advanced bony fishes, that are related to the perchlike fishes common in virtually all marine habitats. This suborder of between 90 and 100 species is primarily confined to the Antarctic region; there it dominates, accounting for an estimated two-thirds of the fish species and 90 percent of the individual fish in the area.

We have concentrated much of our study on two noteworthy adaptations. The first one, clearly crucial to cold-water survival, is the ability to produce compounds that have powerful antifreeze properties. Such compounds depress the freezing point of body fluids. The second adaptation is the development in certain species of "neutral" buoyancy, or weightlessness in water. Weightlessness spares fish from having to expend precious energy on flotation. It appears to have enabled at least two species of Antarctic notothenioids to radiate from the bottom of the sea, where most of them dwell, to the midwaters, which are underutilized.

Several geologic and oceanographic events account for the overwhelming dominance of the Antarctic notothenioids. During most of its history Antarctica was joined to the other southern continents in a great mass known as Gondwanaland. The agglomerate began to break apart perhaps 80 million years ago. The water surrounding Antarctica probably was reasonably warm in the early years, at least in places. In a recent survey of 38-million-year-old fossils found on Seymour Island, one of us (Eastman) and Lance Grande of the Field Museum of

Figure 23 TYPICAL NOTOTHENIOID, *Trematomus nicolai*, lives, feeds and reproduces near the sea floor. The suborder Notothenioidei, a group of bony perchlike fishes, dominates in the Antarctic, accounting for some 90 percent of the individual fish in the region.

Natural History in Chicago found that the coastal waters once supported sharks, saw sharks, rays, ratfish, catfish and other temperate-water groups now absent or poorly represented in the Antarctic.

At about the time the Seymour Island fossil deposit formed, Antarctica became fully separated from Australia and the tip of South America, and its shores were surrounded by vast expanses of cold, deep ocean. Complex environmental changes contributed to the cooling of the Antarctic waters. In the ocean itself one of these changes was the formation of the Antarctic Convergence. This pattern of ocean currents, which lies between 50 and 60 degrees south latitude, surrounded the newly isolated continent and eventually became a formidable thermal barrier, impeding the inflow of warm currents and to a great extent warm-water fishes from the north.

Presumably the cold water caused temperate fishes to disappear—but not the notothenioids. They apparently began to evolve under the influence of the cold. Today, because of both the cold and certain other factors, the Antarctic Ocean has less diversity than even the Arctic Ocean, which supports one and a half times as many species and twice as many families. Among the other factors accounting for this lack of diversity are a paucity of island groups and a deep continental shelf at Ant-

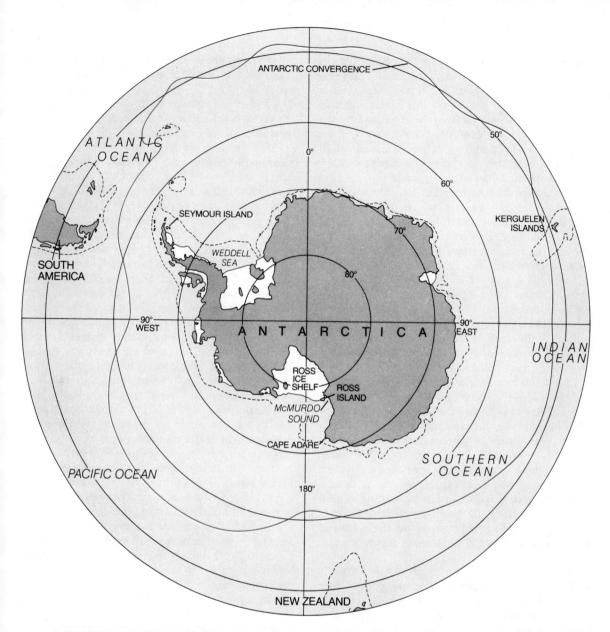

Figure 24 OCEANOGRAPHIC FEATURES that may have influenced the evolution of the notothenioid fishes include a narrow, deep continental shelf and the Antarctic Convergence, a zone of abrupt thermal change (*red line*). The water close to the margin of Antarctica reaches a depth of 1,000 meters (*broken line*), leaving little suitable habitat for many shallow-water fishes that might otherwise compete with the notothenioid fishes. The Antarctic Convergence, which forms the northern boundary of the "Southern Ocean" (where the Atlantic, Pacific and Indian oceans mingle), prevents warm surface water from flowing south into the Antarctic region. The development of the convergence may have contributed to the cooling of the Southern Ocean and hence to the evolution of the notothenioids under the influence of the cold. Today the temperature of the Southern Ocean rarely rises above two degrees Celsius.

arctica's margin; the shallow water that typically surrounds islands and coastal shores is a prime habitat for many fish species elsewhere.

To learn about the nature of the adaptations that have enabled notothenioids to evolve and thrive where other fishes have failed, we and our colleagues periodically travel to Ross Island, about 400 miles south of Cape Adare. For nearly 30 years the National Science Foundation has maintained a biological research station on this small, volcanic strip, which is separated from the mainland by the 40-mile expanse of McMurdo Sound.

One of us (DeVries) began going to the island 20 years ago, primarily to clarify the mechanism by which Antarctic fishes avoid freezing. By that time certain background information had already been uncovered about the conditions in McMurdo Sound. In 1961, for example, Jack L. Littlepage, then at Stanford University, established that the average yearly temperature of McMurdo Sound is −1.87 degrees Celsius, with the range varying only between −1.4 and −2.15 degrees. In the austral summer, from December through February, temperatures rise from −1.9 to −1.8 degrees. Even in summer the water under the ice receives less than 1 percent of the sunlight that strikes the surface, but this is an improvement over the total darkness that prevails for four months of the year.

More hazardous for Antarctic fishes than the dark and the cold is the danger posed by multiple layers of ice. Some two to three meters or more of "annual" sea ice cover the water for at least 10 months of the year—until summer, when storms usually break up the ice and wash it out to sea. An additional meter or two of platelet ice (large, elongated, loosely aggregated crystals) adjoins the undersurface of the annual ice, disappearing starting in mid-December. Part of the year yet another layer of large crystals, called anchor ice, coats the bottom of the sound wherever the water is no deeper than 30 meters.

Ice, as one investigator demonstrated in the 1950's, is a threat because it easily penetrates a fish's gills and integument. Fish, which are cold-blooded and typically have essentially the same temperature as that of their environment, can survive when their blood cools as much as one degree below its equilibrium freezing point: in practical terms, the point at which ice crystals can form. On the other hand, fish can endure such "undercooling" and retain the liquidity of their fluids only if no ice enters the body. When ice is present around a fish that has been undercooled by as little as .1 degree the ice can rapidly propagate across the integument and freeze body fluids. In the presence of ice most tropical and temperate fishes freeze when their body fluids cool to approximately −.8 degree. McMurdo Sound notothenioids, on the other hand, freeze only when their temperature goes down to approximately −2.2 degrees.

With these data in hand, DeVries and his students set about determining the relative contributions of various substances to the freezing-point depression of McMurdo Sound fishes. For most marine fishes, salts (particularly sodium chloride) in the body fluids account for 85 percent of the depression below zero degrees, which is the standard freezing point of pure water. The remainder of the depression can be attributed to small amounts of potassium, calcium, urea, glucose and amino acids, all common constituents of blood and tissue fluids.

In the McMurdo Sound fishes, the workers found that sodium chloride and other ions and small molecules account for only from 40 to 50 percent of the freezing-point depression, even though the concentrations of these substances are somewhat higher than they are in temperate marine species. The balance of the depression, and ultimately the survival of the Antarctic fishes, rests on eight different antifreeze molecules, which are found in the body fluids of most notothenioid species studied to date. The full complement typically appears in most of the body fluids, except in the urine and the ocular fluid and within the cytoplasm of most cells, and it accounts for 3.5 percent of the weight of the fluids.

These antifreeze molecules are glycopeptides. Each consists of repeating units composed of a two-sugar molecule (a disaccharide) covalently bonded to the third amino acid of a peptide chain of three amino acids. The molecules differ primarily in size, ranging in molecular weight from 33,700 daltons down to 2,600 daltons. For clarity every glycopeptide has been assigned a number on the basis of its molecular weight; the largest molecule is No. 1 and the smallest molecule is No. 8. In glycopeptides 1 through 5 the amino acid sequence is alanine-alanine-threonine, and in the other glycopeptides the amino acid proline substitutes for some of the alanines. The antifreeze activity of the eight compounds increases with weight, and all the molecules appear to function similarly.

Notothenioid antifreezes lower the freezing point of fluids in a way that can best be understood by comparing their mode of action with that of more typical body-fluid solutes, such as glucose and sodium chloride. The freezing point of most solutions depends on their "colligative" properties, that is, on the number of solute particles present rather than on the nature of the particles. The more particles there are, the less likely it is that water molecules will aggregate and form an "embryonic" ice crystal. In water, sodium chloride depresses the freezing point nearly twice as much as glucose does because the salt dissociates into separate sodium and chloride ions. Glycopeptide antifreezes, in contrast, act noncolligatively: they can lower the freezing point of body fluids from 200 to 300 times more than would be expected on the basis of particle number. (The melting point also is lowered, but slightly and in a colligative manner.)

By what mechanism might these noncolligative antifreezes keep McMurdo Sound fishes from freezing in ice-laden waters? Chemists have long known that adsorbed impurities can inhibit the growth of small crystals and that, for unclear reasons, impurities consisting of a large number of repeating molecular subunits are particularly effective in this regard. One of us (DeVries) therefore suspected that the glycopeptide antifreezes might safeguard Antarctic notothenioids by adsorbing to minute ice crystals and so preventing their growth. Studies by DeVries and his students John G. Duman and James A. Raymond, then at the Scripps Institution of Oceanography, support this notion. The group found that the glycopeptides do indeed adsorb to ice while it is forming.

Although the events occurring at the molecular level are difficult to view, the group has proposed a likely description. We believe that ice cannot propagate over adsorbed glycopeptide molecules and is forced to grow only in the small spaces between them. Furthermore, these advancing fronts are curved and consequently have a large surface area in relation to their volume. Such ice fronts ultimately lose water molecules to the surrounding liquid, a phenomenon that halts their growth. In order for water molecules to be added to the fronts, the temperature of this surrounding liquid would have

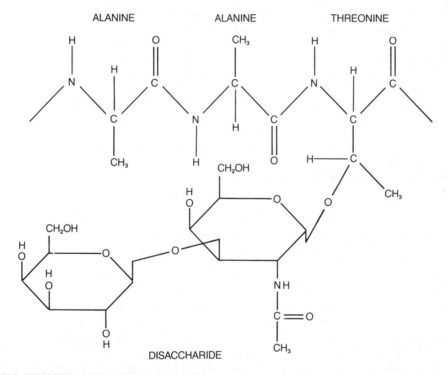

Figure 25 BASIC STRUCTURAL SUBUNIT of notothenioid antifreeze compounds consists of a disaccharide (two-sugar) molecule linked to the third amino acid of a tripeptide (three-amino-acid) chain. The antifreeze molecules, known as glycopeptides, consist of repeating subunits and are identified by number, according to their molecular weight. (No. 1 is the heaviest molecule, at 33,700 daltons; No. 8 is the lightest, at 2,600 daltons.) In glycopeptides 1 through 5 the subunit peptide chain consists of alanine-alanine-threonine, as is shown; in glycopeptides 6 through 8 proline replaces some of the alanines (not shown).

to be lowered. In other words, the development of highly curved fronts depresses the freezing point of the liquid.

Although this hypothesis could be interpreted as indicating that ice crystals form in the body fluids and are then stunted by a barrage of antifreeze molecules, recent experiments suggest otherwise. When notothenioids were placed in ice-free water, they did not freeze until their temperature fell to -6 degrees. Apparently they do not usually develop ice crystals inside the body unless ice first penetrates from the outside. The major threat to the notothenioids' ability to survive the cold must therefore be external ice, and the main role of the glycopeptides is probably to keep the ice from propagating across the integument. Some additional evidence is consistent with this suggestion. When the interior surface of scaleless skin is bathed with a salt solution containing antifreeze molecules, the skin acts as a barrier to ice propagation from the outside. When the antifreeze molecules are not present, ice propagates readily across the integument.

We do not yet know how antifreeze molecules bind to ice, in part because no one fully understands their three-dimensional structure in solution. We do know that the hydroxyl ($-OH$) groups and other polar groups of the antifreezes branch out from the backbone. We also think these groups are binding sites. Indeed, we have shown that the hydroxyl groups in the sugar fraction of the glycopeptides are apparently vital to the antifreeze function of the molecules. When these hydroxyls are experimentally inactivated (by adding an acetyl group, CH_3CO-), the glycopeptide molecules lose their antifreeze effect.

The polar groups can potentially form hydrogen bonds with water molecules in the ice lattice, which are arranged in rows of hexagons that have oxygen atoms at the "corners." For maximal hydrogen bonding to occur between the ice lattice and the polar groups in the glycopeptide, a lattice match would have to be possible; that is, the groups would have to be separated by distances that correspond to the distances between oxygen atoms on a growing front of the ice lattice.

This ideal requirement may be met in reality. Models of glycopeptides reveal that many of the hydroxyls of the sugar side chains are spaced 4.5 angstrom units apart—just about the distance separating some of the oxygen atoms on the horizontal plane of the ice lattice.

Another way antifreezes might bind to ice is through the carbonyl ($-CO-$) groups in the amino acid chain. It could be that the sugar side chains of the glycopeptides somehow keep the polypeptide backbone of the molecule in an expanded conformation. Under this condition every other carbonyl group would project from the same side of the polypeptide and be separated by about 7.3 angstroms. A similar distance separates certain oxygens in the ice lattice.

R egardless of their mechanism of action, the antifreezes of the McMurdo Sound notothenioids are clearly needed all year round. This raises an interesting problem. Antarctic fishes have limited energy stores and must conserve energy, particularly in the austral winter, when the ecosystem's productivity is particularly low. How then do they maintain an adequate supply of antifreezes without expending great energy on synthesis?

The fact that notothenioids have no antifreeze molecules in their urine suggests that some mechanism associated with the kidneys may play a role in conserving antifreezes. In most vertebrates the glycopeptide antifreezes, which are relatively small molecules, would be expected to escape routinely into the urine through the kidney's glomeruli: the tufts of capillaries that act as blood filters. The pressure inside the glomeruli usually forces molecules smaller than 40,000 daltons to pass from the blood into the system of tubules that collect the urine. A fish with glomeruli could theoretically retrieve the antifreezes before the compounds leave the tubules, by breaking the molecules down into their smallest components (amino acids and sugars), returning the components to the blood from the urine and then resynthesizing the glycopeptides. The cost, however, would be high: two energy units for each bond that is formed between two amino acids.

In 1972 DeVries and his student Gary H. Dobbs III, then at the Scripps Institution, examined kidney tissues of notothenioids under a microscope. They found that 12 out of 12 species lack glomeruli. Moreover, studies of antifreeze molecules labeled with radioactive isotopes have demonstrated that the kidneys prevent the glycopeptides from ever entering the urine. These aglomerular species produce urine by a secretory process. Cells that line the walls of tubules draw only selected wastes from the blood, leaving the needed antifreezes in circulation.

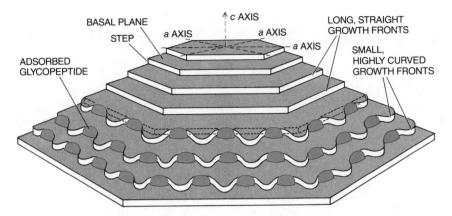

Figure 26 GLYCOPEPTIDES ADSORBED TO ICE probably impede crystal growth. An ice crystal grows when water molecules in the surrounding liquid join at steps on the basal (horizontal) plane of the crystal. In contrast, steps that meet glycopeptide antifreezes divide into many small fronts that become highly curved, a condition that halts their growth if the temperature of the surrounding liquid does not fall.

This eliminates the need for resynthesis and thereby saves energy.

We are not certain that the aglomerular condition of the kidneys in McMurdo Sound notothenioids evolved specifically to conserve antifreezes and energy, but the kidney's importance to those functions is consistent with such a hypothesis. Moreover, we have recently discovered that the New Zealand thornfish (*Bovichthys variegatus*), one of a few notothenioid species found in temperate waters, has many glomeruli. Because the thornfish is among the least specialized of all notothenioids, the fact that it has glomeruli suggests the aglomerular condition of other notothenioids is a specialized adaptation, one that could conceivably have developed specifically as an aid to cold-water survival.

The presence of neutral buoyancy in at least two notothenioid species is another example of an evolutionary adaptation that has enabled certain notothenioids to conserve energy. We discovered this feature about 10 years ago when we captured several specimens of the largest known notothenioid: the Antarctic toothfish (*Dissostichus mawsoni*). (Whereas notothenioids are generally from 15 to 30 centimeters long, the typical toothfish is 127 centimeters long and weighs 28 kilograms.) We had set our fishing line between 300 and 500 meters deep, and we initially assumed the fish rose from their habitat near the sea floor to strike our hooks. We developed some doubts, however, when we noted that *Dissostichus*, as well as *Pleuragramma antarcticum*, the smaller notothenioid species that accounted for 70 percent of the stomach contents of *Dissostichus*, had the streamlined appearance of mid-water fishes. Perhaps *Dissostichus* and its prey were highly modified, mid-water offshoots of the bottom-dwelling notothenioid stock.

In 1978 we began a series of investigations to determine whether these two species might be buoyant enough to live efficiently in middle-level waters and, if they were, which of their adaptations made this buoyancy possible. Neutral buoyancy is actually a comparative measure, derived by dividing a specimen's weight in the water by its weight outside the water and multiplying by 100. The closer the result is to zero, the closer the fish is to being neutrally buoyant. The mean result for *Dissostichus* was .01 and for *Pleuragramma* .6, close enough to zero for these species to be considered neutrally buoyant.

Simple dissection confirmed our belief that these fishes, like their bottom-dwelling relatives, lack a swim bladder, the gas-filled sac that typically results in neutral buoyancy in fish. Something else had to account for our findings. One obvious possibility was a reduction in the bony material of the skeleton, typically the body's densest component. We pursued this idea with the help of an ordinary chef's knife from the mess hall. The knife easily cut through the skull of *Dissostichus*, which consisted largely of cartilage, as did the tail skeleton and the pectoral girdle, another major bone. Cartilage is

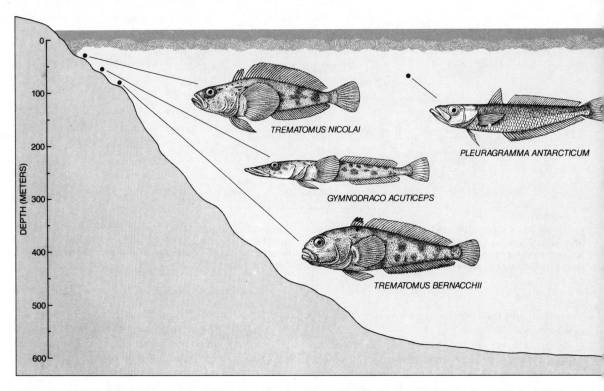

Figure 27 SEVEN SPECIES of notothenioids are among some 14 fish species in McMurdo Sound and some 90 in the Southern Ocean. Many notothenioids, such as *Trematomus nicolai, T. bernacchii, T. loennbergii* and *Gymnodraco acu-ticeps,* dwell on or near the sea floor, albeit at different depths. (The dots indicate typical levels; some fishes have a wider range than is shown.) A few species have radiated from the bottom, notably *Dissostichus mawsoni* and *Pleur-*

considerably less dense than bone, offering a major saving in weight.

To qualify the degree of mineralization of the skeleton, we next employed a furnace to "ash" the skeletons of *Dissostichus, Pleuragramma* and a bottom dweller, *Bovichthys.* This procedure vaporizes organic matter, leaving only the mineral residue of the skeleton. The ashed skeleton constituted .6 percent of the body weight of *Dissostichus* and only .3 percent of the body weight of *Pleuragramma,* in both cases a marked contrast to the 3.8 percent of body weight attributed to the ashed skeleton of *Bovichythys.*

We were not surprised by these results, but we were by our next discovery. An air bubble we had somehow introduced into the vertebral column of a partially dissected *Pleuragamma* specimen moved up and down the interior of the vertebrae when we tilted the fish like a seesaw. In most fishes such movement is not possible: the vertebral col-

umn is solid bone and represents the largest part of the skeleton by weight. Further inspection of *Pleuragramma* revealed that its vertebrae are essentially hollow: a thin collar of bone surrounds a gelatinous embryonic structure (the notochord) that persists into adult life.

In addition to a reduction in bone, an abundance of triglyceride (a type of lipid, or fat, that is less dense than McMurdo Sound seawater) would also contribute to buoyancy. Both *Dissostichus* and *Pleuragramma* have triglyceride deposits, although the deposits take different forms.

A cross section of *Dissostichus* glistens with lipid that fills fat cells in two major deposits: a blubber-like, two- to eight-millimeter thick layer under the skin (accounting for 4.7 percent of the body weight) and a more dispersed deposit throughout the muscle fibers of the trunk (accounting for 4.8 percent of the body weight). *Pleuragramma,* in contrast, has a unique method of lipid retention for a vertebrate: it stores its lipid in sacs rather than in cells. These sacs

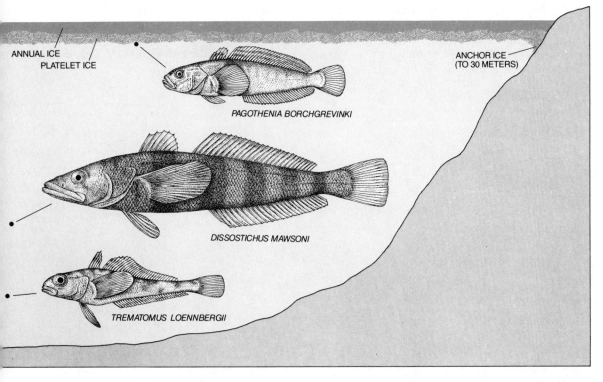

ANNUAL ICE
PLATELET ICE
ANCHOR ICE
(TO 30 METERS)

PAGOTHENIA BORCHGREVINKI

DISSOSTICHUS MAWSONI

TREMATOMUS LOENNBERGII

agramma antarcticum, which are mid-water fishes that live at depths ranging from the ones shown to 500 meters, and *Pagothenia borchgrevinki*, which is adapted to life in and under the platelet ice. *Dissostichus*, shown as larger than the other fishes, is six times as long and 250 times as heavy as most notothenioids.

are from .2 millimeter to three millimeters in diameter and are abundant under the skin in the pectoral region, near the center of gravity. The largest sacs, however, are found between the muscles, deeper in the body.

We have not yet fully determined the relative benefits of storing lipid in sacs as opposed to in fat cells. We originally thought that lipid sacs were exclusively designed for buoyancy and that their lipid stores were unavailable as an energy source; *Pleuragramma* has few fat cells, and such cells would seem to be important for regulating lipid removal. More recently we have found that muscle cells adjacent to the sacs have large vacuoles, or membrane-bound cavities, that may contain triglyceride. Perhaps the muscles add lipid to the sacs for buoyancy and also remove them when they are needed for energy. We intend to test this hypothesis on our next trip to the McMurdo Sound research station.

Skeletal reduction and lipid deposition together can account for the reduced density and the neutral buoyancy of both *Dissostichus* and *Pleuragramma*. In turn weightlessness certainly facilitates their exploitation of the underutilized mid-waters of the Antarctic. Indeed, both species are abundant and widely distributed there. *Pleuragramma*, for instance, is known to be the dominant species in the Ross and Weddell seas, as well as in the waters of McMurdo Sound.

The adaptations that have enabled notothenioids to radiate into the mid-waters fascinate us from an anatomical point of view; we are also struck by their ecological implications. It is time to revise the standard accounts of the mid-water food web. As it is usually described, the web is short, including only planktonic plant life, krill (a shrimplike crustacean), seals and whales. We and other investigators have shown, however, that the prevailing vision of the composition and complexity of the food web should be expanded to include neutrally buoyant notothenioids and perhaps other organisms as well. Indeed, in areas of the Antarctic Ocean where krill is sparse *Pleuragramma* may take the place of krill in the food chain. Finally, from yet another point of view, we hope our studies of buoyancy and anti-freezes provide a glimpse of the extraordinarily wide scope of adaptation and evolution that has occurred at the coldest extreme of the environmental spectrum, a place that was once widely thought to be incompatible with significant marine life.

Intertidal Fishes

Fishes that live between the tides are alternately buffeted by waves and isolated in pools and on mud flats. Anatomy, physiology and behavior suit them to their rigorous habitat.

. . .

Michael H. Horn and Robin N. Gibson

January, 1988

The intertidal zone—the band of shoreline that is between the low-tide mark and the high-tide mark—is a demanding environment. Roughly twice a day, at low tide, the intertidal zone is cut off from the open ocean. The water that remains is left in isolated tide pools or under rocks, or else it has combined with the substratum to form mud flats. When the water returns at high tide, the intertidal zone is submerged to rejoin the broader ecosystem of the ocean. Any animal that lives in the intertidal zone must be able to spend much of its time either completely out of the water or at least partly exposed to the air. The periods during which it can feed are set by the tidal cycle, and it must withstand wide variations in water chemistry and the nearly constant wave action and turbulence that are characteristic of the zone.

The fixed and slow-moving animals inhabiting the intertidal zone, such as barnacles, limpets and periwinkles, have been studied intensively, primarily because they are abundant and accessible. Intertidal fishes have been more difficult to study, in part because of their mobility and in part because they are generally well hidden. Many of them are cam-

ouflaged by protective coloration, and at low tide many species live under rocks or within clumps of vegetation (although some species, such as the tropical mudskippers, can live in the open on mud flats).

Nevertheless, in the past 15 to 20 years our knowledge of these fishes has increased to the point where it is now possible to give a fairly coherent picture of how they make a living in this unusual habitat. Our work and studies by other investigators have revealed that intertidal fishes are remarkably well suited to the environment in which they live; they are not simply ocean animals that have been stranded by the ebbing tide. Indeed, these fishes have evolved to become integral components of the intertidal ecosystem, with each species of fish occupying a characteristic vertical distribution and type of habitat on the shore.

Many intertidal fishes display the specialized anatomical features characteristic of fishes dwelling on the bottom of shallow, turbulent waters. The most obvious of these features is small size. Intertidal fishes are rarely longer than 30 centimeters, and most of them are less than 20 centimeters long. Their small size enables them to occupy holes, crevices and spaces under rocks, and it reduces the risk of their being swept away by surge or waves. The ability of many intertidal fishes to live in confined spaces is also enhanced by such characteristics as very thin bodies (in the gunnels and pricklebacks) or flattened, horizontal profiles (in the sculpins and clingfishes).

In addition to specialized body shapes and sizes, many intertidal fishes have distinctive fin structures. For example, the highly terrestrial mudskippers can raise themselves on their paired pectoral fins when they are moving about on the land. In some intertidal fishes (such as clingfishes, gobies and snailfishes) the pelvic fins are modified to form suction cups by which the fishes can attach themselves firmly to the substratum. On the other hand, in the fishes that are most highly modified for dwelling in crevices or holes (such as certain gunnels and pricklebacks) the pectoral and pelvic fins are greatly reduced and the dorsal and anal fins are long, low and often united with the tail fin.

The skin of intertidal fishes is generally tough, and so it can withstand repeated scraping against

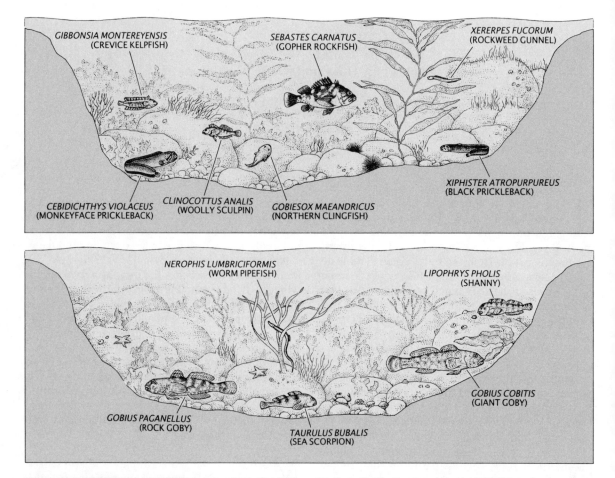

Figure 28 INTERTIDAL FISHES common along the California coast (*top*) and the Atlantic coast of France (*bottom*) are shown in idealized tide pools at high tide, when they are active. At low tide, when the water level in the tide pools falls or they dry out entirely, the fishes are usually hidden in seaweed and under rocks. Although each site is host to a distinctive set of species, the fishes share anatomical features that suit them for hiding in tight spaces and withstanding wave action, among them small size, thin bodies or flattened profiles, and pelvic fins (the paired fins on their underside) reduced in size or modified for clinging to the bottom.

the substratum. Some of the fishes (such as blennies and clingfishes) have no scales; in others (such as gunnels and pricklebacks) the scales are greatly reduced and in still others (the gobies) they are attached very firmly. Many of the fishes secrete large amounts of mucus, which may provide lubrication in confined spaces. (It may also help to reduce the loss of water when the fishes are exposed to the air.)

A standard feature of intertidal fishes is cryptic coloration. Flatfishes on sandy shores (plaice and the California halibut, for example) are renowned for their remarkable ability to match their pigmentation with the colors and patterns of the bottom. On rocky shores, where the background colors can be more varied, intertidal fishes exhibit a broad range of colors and patterns. Species that live within algal vegetation usually have striking colors that match the surrounding seaweed. For instance, the gunnels *Apodichthys flavidus* and *Xererpes fucorum* can range in color from tan to bright green to deep red, depending on the color of the algae in which they live. These fishes, which change color slowly, apparently gain pigments from their invertebrate prey, which also inhabit seaweeds. Other fishes can change color within seconds to blend in with sessile invertebrates or encrusted rock surfaces.

Many intertidal fishes have negative buoyancy (they are heavier than seawater), which enables them to lie effortlessly on the bottom, where cover is nearby and water velocities are lowest. Such fishes have either no swim bladder (the gas-filled organ characteristic of most bony fishes) or a greatly reduced one. The relatively high specific gravity of intertidal fishes helps to explain their labored swimming, which usually consists of brief excursions from cover or short darts from one place to another.

The intertidal region experiences frequent and dramatic fluctuations in environmental conditions, threatening its inhabitants with desiccation during exposure to air and confronting them with variations in oxygen availability, salinity and temperature that would be lethal to other species. Resident fishes must be behaviorally and physiologically equipped to cope with such great changes. Even intertidal species vary in their tolerance to these stresses, particularly to desiccation. The variations help to determine the vertical distribution of species on the shore. (See Figure 30.)

Several studies have shown that many intertidal fishes can tolerate considerable loss of water. For example, William H. Eger, while a graduate student at the University of Arizona, found that certain clingfishes from the Gulf of California can survive for as long as 93 hours out of the water if the humidity is high (90 percent) and can sustain water losses as great as 60 percent of their total water content in environments where the humidity is low (5 percent). This is a higher loss of water than can be tolerated by most of the amphibians that have been studied so far. (In fact, certain intertidal fishes, the tropical mudskippers, virtually *are* amphibians: unlike many other intertidal species, they are often active out of water and may spend between 80 and 90 percent of their time on land.)

The rate of desiccation may be slowed in some species by such anatomical features as a thickened epidermis and the presence of mucus-secreting cells in the skin, but intertidal fishes apparently have no other physiological mechanisms for reducing or preventing water loss. Indeed, live and dead fish exposed to the air lose weight at about the same rate. It seems, then, that behavior—simply trying to avoid desiccating conditions—plays an important role in survival. Most species remain inactive when they are exposed at low tide. Fishes that are active out of water, such as the mudskippers and rockskippers (which are not related to each other), ensure against excessive drying of the skin and respiratory surfaces by returning frequently to the water or remaining within reach of waves or spray.

Changes in the supply of oxygen cause two kinds of problem for intertidal fishes. First of all, the fishes must somehow obtain oxygen even when they are out of the water for several hours. The availability of oxygen is not the problem, of course, because the concentration of oxygen is much higher in air than in saturated water. The difficulty comes in acquiring the atmospheric oxygen. In general, gills tend to collapse in the air; their thin, flexible structure is much better suited to operating under water. The problem is solved in mudskippers and rockskippers by shorter, thicker gill filaments, which prevent collapse of the gills and enable them to function for breathing air. Most species that can breathe air have other specialized features, such as increased numbers of blood vessels in the skin and in the lining of the mouth and pharynx.

In these air-breathing species the rate of respiration in air is nearly equal to the respiration rate in water. Furthermore, according to Christopher R. Bridges of the University of Düsseldorf, the concen-

trations of lactate in the blood and muscle of inter-tidal species show that the rate of anaerobic metab-olism (metabolism that does not directly involve oxygen) does not increase when the fishes are ex-posed to the air, indicating that the supply of oxy-gen to the fishes' cells is not greatly reduced when they must breathe air. The fishes can therefore function out of water without any major changes in their mechanisms of metabolism.

The second kind of respiratory problem confronts intertidal fishes that inhabit isolated tide pools high above the low-tide mark. Photosynthesis by sea-weeds in the pools and respiration by the resident plants and animals cause strong diurnal fluctuations in the amounts of dissolved oxygen and carbon dioxide in the pool water. During the day seaweed photosynthesis can supply amounts of oxygen ex-ceeding the respiratory demands of a pool's inhabi-tants, resulting in high levels of oxygen and low levels of carbon dioxide. At night, when photosyn-thesis ceases, respiration by the pool's plants and animals depletes the oxygen supply and increases the level of carbon dioxide.

During the day tide-pool fishes do not seem to change their rate of oxygen consumption in re-sponse to prevailing conditions. At night, however, some species do reduce their oxygen consumption to make up for the poor availability of oxygen. When oxygen reaches critically low levels, some species have been observed to ventilate their gills with the thin film of water at the surface of the pool, where there is more dissolved oxygen; in the laboratory, fishes in water that has an extremely low level of dissolved oxygen sometimes emerge from the water and climb onto exposed surfaces.

Along with unusual anatomical and physiologi-cal traits, fishes in the intertidal zone have distinctive behavior patterns. For example, the re-productive process of most intertidal fishes follows a generalized pattern. First the male selects a

spawning site in a sheltered location (under a stone or within vegetation), which in many species is within a territory the male defends before and after spawning. Then he attracts one or more females to the site; each female deposits a single batch of eggs on the substratum. The male fertilizes the eggs and then usually tends them until they hatch.

The male's need to choose and protect a cryptic spawning site puts a number of constraints on his courtship behavior. He cannot leave the site to search for mates, and so he must attract females to him. In order to attract females the male relies on numerous displays. For example, some male blen-nies advertise themselves by performing vertical loops, and they induce the female to enter the spawning site by vigorous head movements. The ability of the male to gain attention by displays is augmented by tentacles, crests and distinctive color patterns. Certain species also utilize pheromones, or specific olfactory signals, for communication be-tween the sexes.

Parental care of the eggs is the rule, whether the eggs are attached to the substratum (as they are in most species) or are formed by the parent into a cohesive but unattached mass. Usually it is the male who guards the eggs, although in some species it is the female and in others it is both parents. Parental attention safeguards the eggs from predators and probably keeps them from being swept away by surge and wave action.

After hatching, the larvae of most intertidal fishes develop in the ocean itself, as members of the plankton. How they return to the intertidal zone has remained largely a mystery. One possible answer involves the slicks (long strips of smooth water run-ning roughly parallel to the shore) that are produced by the action of internal, or subsurface, waves in the ocean. Recent research has shown that slicks con-tain higher concentrations of fish larvae than adja-cent rippled areas do. Slicks are common coastal features in regions where the temperature or the salinity of the water column changes sharply with depth, and they generally move slowly toward the shore. Perhaps they provide a mechanism by which the larvae of some intertidal fishes can return to the intertidal zone.

On the other hand, work by Jeffrey B. Marliave of the Vancouver Public Aquarium suggests that the larvae of some species may not need to be brought back from more distant waters. He found that lar-vae of some species living in rocky areas remain close to the shore, resisting any forces that might disperse them offshore or even along it. In these

Figure 29 CRYPTIC COLORATION conceals a sculpin, a species of *Clinocottus* (*top*), and a gunnel, *Xererpes fucorum* (*bottom*), against their backgrounds, respectively a sponge and a clump of seaweed. Such camouflage, which probably affords protection against predators, is common among tide-pool species. The sculpin can change color to match its surroundings by means of specialized pigment cells; the gunnel apparently takes its green color from invertebrates it preys on. The sculpin was photographed in its habitat by Anne Wertheim, and the gunnel—a captive specimen—by one of the authors (Horn).

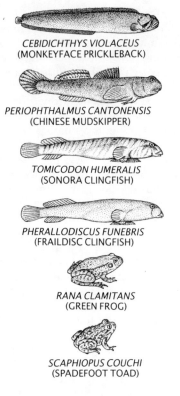

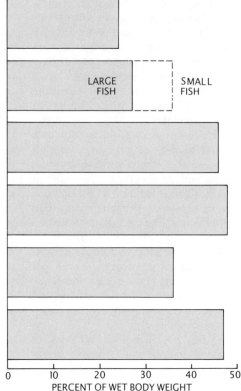

CEBIDICHTHYS VIOLACEUS
(MONKEYFACE PRICKLEBACK)

PERIOPHTHALMUS CANTONENSIS
(CHINESE MUDSKIPPER)

TOMICODON HUMERALIS
(SONORA CLINGFISH)

PHERALLODISCUS FUNEBRIS
(FRAILDISC CLINGFISH)

RANA CLAMITANS
(GREEN FROG)

SCAPHIOPUS COUCHI
(SPADEFOOT TOAD)

LARGE FISH   SMALL FISH

PERCENT OF WET BODY WEIGHT

Figure 30 TOLERANCE TO DESICCATION of some intertidal fishes is comparable to that of amphibians. The bar graph shows the percent of normal weight to which each species could be reduced without killing the animal. The ability to survive water loss is one factor determining how high on the shore a species is found and its level of activity at low tide. Mudskippers, for example, are able to crawl on exposed mud flats and breathe air.

species individuals from different regions would have little opportunity to mingle, and so even populations that are relatively close to one another on the coastline might be genetically isolated.

Another common feature of many intertidal fishes is their homing ability. Intertidal fishes must undertake periodic foraging excursions, and as a result they risk being stranded in unfavorable locations at low tide. Many species have developed homing abilities that enable them to return to a particular location, generally a tide pool, that provides suitable refuge. (Species that do not need to find tide pools but instead live mainly under rocks or in vegetation generally do not have such homing ability.) The ability to home and to become familiar with the topography of a particular area may also help intertidal fishes to escape predators and turbulent weather at high tide.

Most studies of homing behavior done so far have involved the tidepool sculpin *Oligocottus maculosus*. A standard technique is to label fish captured in a given area by attaching colored plastic beads to them and then move the marked fish to a different location. The investigator later revisits the site from which the fish were originally taken to see how many can be recaptured there. John M. Green of Memorial University of Newfoundland and Gwenneth J. S. Craik, then at the University of British Columbia, found that as many as 80 percent of *O. maculosus* return to their native pool after having been displaced by distances as great as 100 meters. Young fish had poorer homing ability than older ones and were unable to retain their homing ability for as long. Some older fish could still return home after having been kept in the laboratory for as long as six months before being released.

Both vision and olfaction appear to be important

for homing in the tide-pool sculpin; it remains to be determined whether one sense is more important than the other. Craik has concluded that either sense can be sufficient for older fish, whereas both of them are necessary for younger specimens.

D efinitive studies on the details and significance of homing have yet to be done, but new techniques based on ultrasonic telemetry are finally making such research possible. In a recently completed study telemetry enabled one of us (Horn), working with Scott L. Ralston of Deep Ocean Work Systems in San Pedro, Calif., to track the movements of several monkeyface pricklebacks (*Cebidichthys violaceus*) continuously for periods of up to two weeks. A small transmitter attached to the inside surface of each fish's gill cover emitted high-frequency sounds, which were detected by three hydrophones deployed on the sea floor. The hydrophones relayed the signals to an onshore receiver and microcomputer, which determined the fishes'

locations, displayed them on a screen and stored the information for later analysis.

We found no evidence of homing; our results suggest instead that the monkeyface prickleback stays within a home range of no more than a few square meters of intertidal habitat. The species also proved to be sluggish: each individual we tracked was active for a total of only about five minutes a day, mostly during the incoming tide. We look forward to applying the technique to other species. Increasing miniaturization of transmitter components should soon make it possible to track fishes even smaller than the pricklebacks we followed, which ranged from 23 to 41 centimeters in length.

It ought to be advantageous for intertidal fishes to sense the state and motion of the tide, in order to synchronize their activity with its regular, twice-daily rhythm. Conceivably the fishes could do so by responding directly to such signals as changes in temperature, salinity, turbulence or light intensity. Yet there is another mechanism: recent studies by

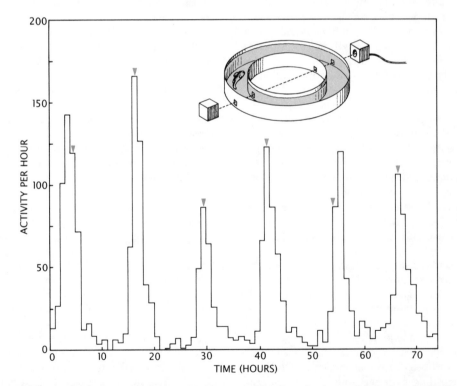

Figure 31 **INTERNAL ACTIVITY RHYTHM of the shanny** (*Lipophrys pholis*), measured in the laboratory, shows regular peaks matching the expected times of high tide (*triangles*) in the fish's natural environment. The rhythm — characteristic of many intertidal species — was recorded under constant conditions in an apparatus that confined the fish in a circular trough (*top*); a photocell registered an activity "count" each time the fish interrupted a beam of infrared light. The internal clock governing such activity rhythms may be set by changes in water pressure that accompany the rise and fall of the tide.

one of us (Gibson) have shown that even when such potential indicators are held constant in the laboratory, some fishes, from a variety of families, still display rhythms of behavior that are related to the tidal period. They alternate periods of rest at the expected time of local low tide with bouts of activity at or near the expected time of high tide.

We do not know exactly how such an endogenous rhythm comes to be established, although laboratory investigations show that the changes in hydrostatic pressure associated with the rise and fall of the tide are at least partly responsible. Many intertidal fishes have no swim bladder, the organ usually considered to be responsible for the detection of pressure by bony fishes, and so how they detect these slow pressure changes remains a puzzle.

In nature the fishes' responses to the tides are sometimes modulated by daily cycles of light and dark, and perhaps also by an inherent circadian (approximately 24-hour) rhythm. Field studies have demonstrated that many species are strongly diurnal in their patterns of activity. This is probably related to their dependence on vision for such major activities as foraging, mating and avoiding predators. Knowledge of daily rhythms in intertidal fishes, like knowledge of their homing behavior, will be greatly increased by studies that exploit ultrasonic telemetry.

In addition to its other characteristics, the intertidal zone presents a particularly rich area for feeding. The great mass and diversity of living material in the intertidal zone, particularly on rocky shores in temperate waters, is matched by the variety of feeding styles found there. The complex interrelations of the intertidal food chain are further complicated because many species change their eating habits seasonally or annually and still others change their tastes as they age.

Most intertidal fish species are carnivorous, but some eat a combination of plant and animal foods and a few are strictly herbivorous. The intertidal fishes that live in temperate climates and eat primarily seaweeds are particularly interesting. How do they extract an adequate diet from a food source that is so relatively low in nutrients (particularly protein) and whose cell walls consist of cellulose, a largely indigestible material? Indeed, some marine biologists have suggested that these fishes may not digest seaweed tissue at all but rather may feed on small organisms on the seaweed's surface.

Some insight has been gained into the feeding habits of these fishes from observations of the monkeyface prickleback and another California prickleback, *Xiphister mucosus*, by one of us (Horn) with Steven N. Murray and Margaret A. Neighbors of the California State University at Fullerton. These fishes begin life as carnivores, but when they are less than a year old, they become complete herbivores, feeding on certain species of red and green seaweeds, particularly those that are relatively rich in protein or carbohydrate. While he was a graduate student at Fullerton, Kevan A. F. Urquhart demonstrated that the monkeyface prickleback digests the seaweed not by breaking down its cell walls with specialized enzymes but by retaining the food in its gut for a long time (about 50 hours) and gradually leaching carbohydrate and protein from the seaweed with acidic stomach fluids.

Little is known about the impact of intertidal fishes as predators on the plant and animal populations of rocky shores. Limited evidence suggests that in such places as the Great Barrier Reef fishes that browse in the intertidal zone are the major factor controlling the abundance and composition of the intertidal biota. Investigation of the rates at which intertidal fishes of California's rocky coast consume various kinds of prey has led Gary D. Grossman of the University of Georgia to propose that these fishes exert strong selective pressures on many species of intertidal algae and invertebrates. Progress in this area of investigation has been slow because of the difficulty of manipulating (or even determining) the numbers and distributions of intertidal fishes.

The evolutionary forces underlying the colonization of the intertidal zone by fishes are far from being completely understood. These species are able to cope adequately with the demands of their habitat: they tolerate its dehydrating conditions, withstand its characteristically extreme changes in water chemistry, breathe air when necessary, feed on a wide variety of food resources and guard their eggs against both physical and biological dangers.

But if the invasion of the intertidal zone by fishes is to be understood in the context of natural selection, one must ask what the advantages are of living there. It has not been possible to test whether the intertidal zone offers lower levels of competition and predation than exist in the adjoining subtidal zone. It is often suggested that the main advantage

Figure 32 ROCKY COAST, with its tide pools, boulders and seaweeds, offers a diversity of habitats for intertidal fishes. It is also a turbulent and dynamic environment: the fishes that make their home there must cope with waves, surge and the physiological stresses imposed by the ebb and flow of the tide. The photograph was made on Washington's Olympic Peninsula by Anne Wertheim.

of living in the intertidal zone is the opportunity to escape from aquatic predators, but at low tide, when the danger from aquatic predators is least, intertidal fishes may become vulnerable to terrestrial and aerial predators. The relative advantages of the intertidal zone and the impacts of different kinds of predators are difficult but significant areas of study that can be explored only by means of carefully conducted field experiments.

SECTION

III

BEATING THE ODDS

. . .

Introduction

...

One advantage of exploiting a seemingly impossible niche, which is the essence of life on the edge, is that there is usually little competition from other species. An insect niche in winter, for instance, when nearly all the occupants are dormant, is an open invitation to any creature hardy enough to take advantage of it. By flying when insectivorous birds have migrated to warmer climes and bats are in hibernation, the winter moths described in Chapter 6, "Thermoregulation in Winter Moths," gain the additional advantage of having no predators to worry about.

Among the rest of the Lepidoptera, however, predation has been perhaps the most important selective factor. Most moths fly at dusk or during the night, and several have evolved ears with which to listen for the ultrasonic sonar of the bats that hunt them. These ears allow the moths to dive for cover when a bat approaches [see "Moths and Ultrasound," by Kenneth Roeder; SCIENTIFIC AMERICAN, April, 1965].

The winter moths' trick of being active during the coldest part of the year brings its own set of problems, though—staying warm enough to fly in search of food and mates, and not freezing when stationary. In fact, it may have been the bat-detecting ears, superfluous during the winter, that provided a key preadaptation toward allowing winter moths to conserve heat. Combined with a replumbing of their circulatory system, their insulation has made the moths so good at retaining heat in their muscle-rich thoraces that they risk cooking themselves from within at milder temperatures— temperatures barely warm enough to allow conventional moths to be active.

Another niche ripe for exploitation is that created by the many plants that are toxic to most predators. The familiar monarch butterfly, for example, thrives because its biochemistry has evolved to allow it to survive the cardiac glycosides that milkweed produces to discourage herbivores. Milkweed pays a price for the privilege of being toxic in terms of slower growth, since much of its energy goes to synthesize poisons and sequester them where they

will not damage the plant itself. Monarch caterpillars are less efficient organisms as well because of all the biochemical effort that goes into dealing with the same poisons. The monarch recoups some of this investment, however, by concentrating the toxic chemicals it ingested as a caterpillar in its wings and other vulnerable spots. Birds associate its bright patterns with illness and avoid it. [See "Milkweeds and their Visitors," by Douglas H. Morse; SCIENTIFIC AMERICAN, July, 1985; and "Ecological Chemistry," by Lincoln Brower; SCIENTIFIC AMERICAN, February, 1968; Offprint 1133. Another such interaction is described in "A Seed-Eating Beetle's Adaptations to a Poisonous Seed," by Gerald A. Rosenthal; SCIENTIFIC AMERICAN, November, 1983.]

The practice of synthesizing poisons to ward off herbivores is widespread in plants, and we owe many of our drugs to these efforts: nicotine, cocaine, caffeine and the cannabin of marijuana are each chemicals toxic or unpleasant to most animals [see "The Chemical Defenses of Higher Plants," by Gerald A. Rosenthal; SCIENTIFIC AMERICAN, January, 1986]. A variety of mammals have evolved to ingest the inedible, however, and in Chapter 7, "The Physiology of the Koala," we see how the poisonous leaves of the eucalyptus provide sustenance for the specialist koala. Even with its ability to detoxify eucalyptus oil, the koala is on the edge. Since it lacks any fat deposits for storing energy, this marsupial lives literally from hand to mouth. It is also in constant danger of running out of nitrogen, particularly when leaves are tough and indigestible. The koala's system tends to excrete the indigestible bits before it can gain access to the nitrogen.

Mammals (and even the infamous termite) are unable to digest the cellulose walls of plant cells and depend instead on symbiotic microorganisms in the gut, which digest the cellulose and are digested in turn. The tougher and drier the eucalyptus leaves, the longer the koala must leave them in the digestive system to ferment. It is even possible for a koala to starve with a full stomach. What chance preadaptation allowed the ungulates (antelope and cattle) to evolve a huge rumen where plant matter ferments *before* it passes to the stomach is a mystery; other herbivorous animals like the koala and the horse must depend on a less efficient "afterburner" strategy.

The nitrogen crisis that the koala's inefficient digestive system creates is paralleled in the plant world as well. Since nitrogen cannot be fixed directly from the air in the presence of oxygen, it is of necessity restricted to anaerobes (or, in the special case of cyanobacteria, to heterocysts). Plants such as clover and alfalfa have evolved a symbiotic relationship with anaerobic nitrogen-fixing bacteria, which are hospitably accommodated in special root nodules. Some plants depend on nitrogen that has been recycled from other plants or plant parts that have died and enriched the soil. Animals return plant nitrogen eventually in the form of waste products or decaying corpses, but despite this the nitrogen supply is often limiting.

In the nitrogen-poor bog habitat natural selection has led to a complete turning of the tables: many plants eat animals to obtain nitrogen (and phosphorus as well). In Chapter 8, "Carnivorous Plants," we see the variety of tricks that have evolved for capturing and consuming prey, including a number of adaptations that intriguingly parallel the physiology of animals. The rapid closing of the Venus's flytrap is based on the same strategy for movement used by earthworms and other creatures with hydrostatic skeletons, and the highly infolded absorptive surfaces of plant stomachs could readily be mistaken for the villi of animal intestines, which serve the same function.

Thermoregulation in Winter Moths

Curiously lacking in highly specialized adaptations for the cold, certain nondescript moth species can nevertheless do what their relatives cannot: fly, feed and mate at near-freezing temperatures.

. . .

Bernd Heinrich

March, 1987

Winter means death to the adults of many insect species in the middle and upper latitudes of the north. If the cold does not kill them, food shortages often do. Ironically, the same challenges that doom most insects make the northern landscape a refuge for others. Winter's trials force birds and bats, the animals' major predators, to fly south or (in the case of certain bats) to hibernate. The cold even kills some parasites.

Among the few insects that have managed to adapt to a winter existence are some 50 species within the Cuculiinae subfamily of the widespread Noctuidae, or "owlet moth," family. The cuculiinids are dull-colored night fliers populating northern deciduous forests. Their winter-adapted varieties have somehow reversed the typical life cycle of the noctuids.

Most owlet moths are active only on warm summer nights. They die off as the winter approaches, leaving behind eggs, larvae (caterpillars) or pupae that remain inactive until the spring. The winter noctuids, in contrast, emerge as adults in the fall or late winter, when they feed, mate and lay their eggs before dying in the spring. Their caterpillars feed in the early spring (eating the buds of forest trees) and then are quiescent throughout the summer. (Adult winter moths generally feed on the sap of injured trees, although on late-fall nights a few years ago I saw many of them feast on the blossoms of witch hazel, Vermont's latest-blooming plant. Until that time no one knew just how the plant was pollinated.)

How do the winter moths survive when other moths die? What enables them to avoid freezing as they rest, and what makes it possible for them to fly—and so to seek food and mates—in the cold? I tackled the latter question first because I was intrigued by an apparent contradiction: I was sure these moths, like other moths that fly, were endothermic, or able to produce heat by their own metabolism. I also assumed they required high body temperatures in order to fly. Nevertheless, it seemed unlikely that endothermy alone could enable the animals to attain the elevated temperatures needed for flight on cold and sometimes snowy nights.

I suspected the animals had a high flight temperature because summer moths and tropical moths that have a similar body size and wingbeat fre-

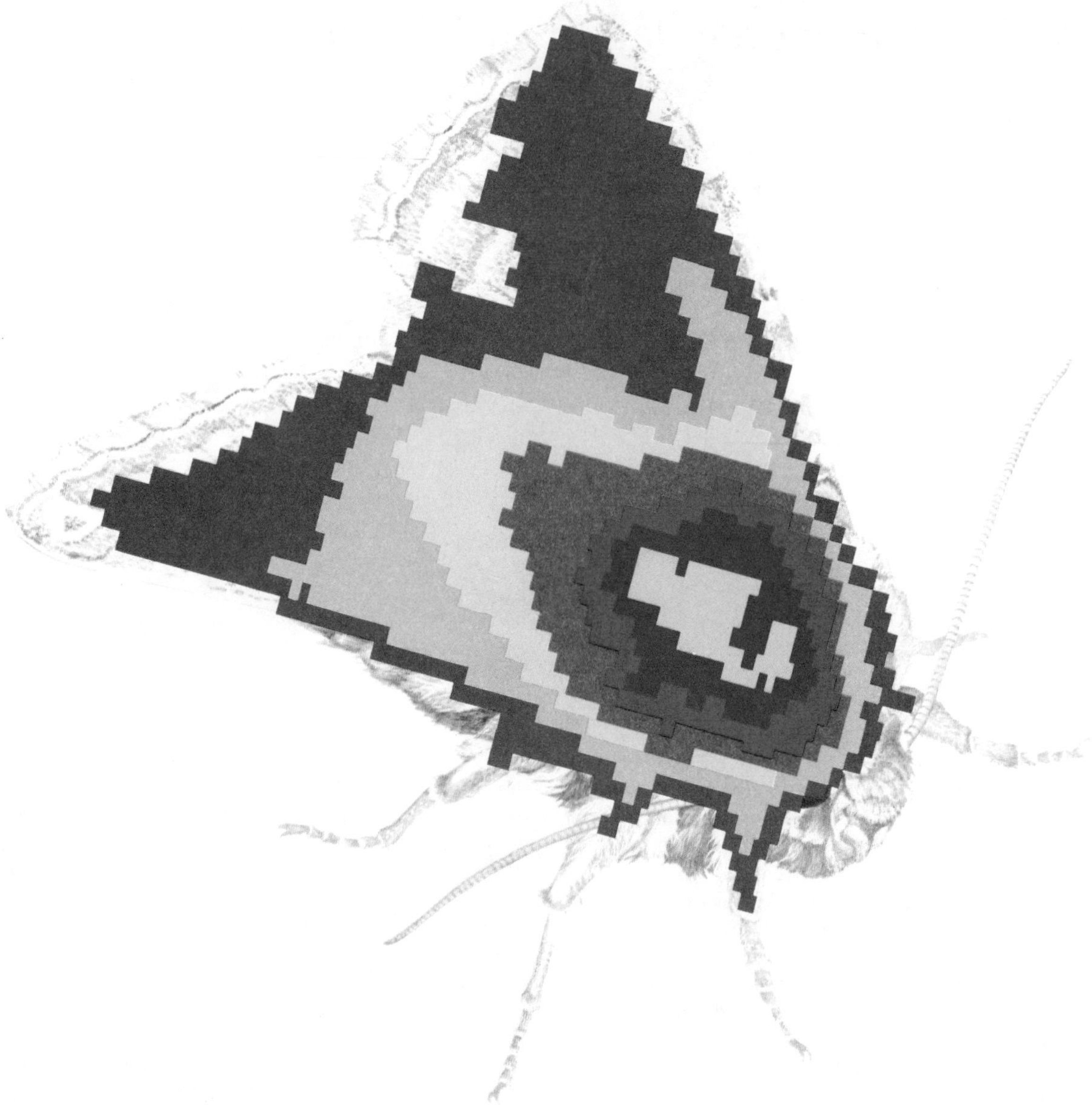

Figure 33 WINTER MOTH of the Noctuidae family is portrayed with a superimposed infrared map to show the distribution of heat immediately after flight. Yellow indicates the highest temperature, followed in sequence by red, pink, dark blue, green, light blue and purple. In order to fly, the winter noctuids must keep their wing muscles (*yellow, red and pink regions*) heated to about 30 degrees Celsius.

quency can fly only when the thorax, which contains the flight muscles, reaches at least 30 degrees Celsius (86 degrees Fahrenheit). On the other hand, different data suggested that the tiny insects (weighing less than .2 gram) would cool too quickly to maintain such temperatures. Small animals, which have a high ratio of surface area to volume, cool faster than larger ones and have greater difficulty retaining heat. Indeed, for many years it was thought that bats, shrews, and hummingbirds were the smallest endothermic animals. The minutest of these vertebrates weigh three grams; they are still virtual giants compared with the cuculiinids.

To resolve the matter I first had to capture winter moths, which I did by smearing tree trunks with bait: diluted honey, maple syrup, beer, or various other sweet concoctions. Temperature measurements obtained with hair-thin thermocouple probes revealed that my initial suspicion was correct: the insects do require—and generate—a high thoracic temperature for flight. Although they have the same temperature as the air when they rest, they endothermically heat themselves to 30 degrees C. or more before flying, even in air temperatures near zero (the freezing point of water).

Having established that the moths generate their own heat, I attempted to determine whether they have any special physiological tricks for doing so. It was obvious from their vibrating wings that the insects obtain heat by shivering, but this was not in itself special. Ann E. Kammer, now at Arizona State University at Tempe, had shown earlier that Lepidoptera (moths and butterflies) shiver during preflight warm-up by simultaneously contracting their major upstroke and downstroke muscles. Yet something about the cuculiinid behavior was striking:

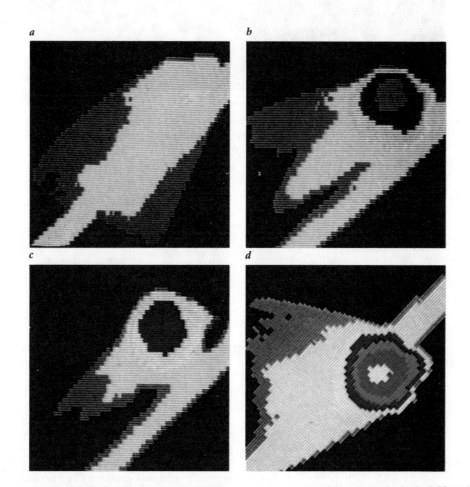

Figure 34 INFRARED PHOTOGRAPHS of *Eupsilia* capture the moth in successive stages of warming up for flight. Like many moths, the cuculiinids heat up by shivering: simultaneously contracting their upstroke and downstroke muscles. In *a* and *d* the moth, perched on a stick, is seen from above, in *b* and *c* from the side. The thorax is the round area that first appears in *b* (*dark blue*). As in Figure 33, yellow represents from 26.6 to 30.9 degrees C.; red, 24.8 to 26.5; pink, 22.4 to 24.7; dark blue, 19.6 to 22.3; green, 17.2 to 19.5; light blue, 14.0 to 17.1; and purple, 11.0 to 13.9 degrees.

Figure 35 TWO WINTER NOCTUIDS, *Eupsilia* **(***left***) and** *Lithophane* **(***right***), eat honey spread on a tree trunk as bait. Some 50 noctuid species, all from the Cuculiinae sub-family, are active in northern winters. The winter cucu-** liinids resemble many summer noctuids anatomically but can warm their flight muscles at lower air temperatures. The cuculiinids are also better able to retain heat in their thorax, the region that houses the wing muscles.

some of them began to shiver at much lower temperatures than other moths of the same size.

Winter cuculiinids generally become active only when the air temperature exceeds zero degrees C., but sometimes they initiate shivering at temperatures as low as −2 degrees. No other moths are known to shiver until the air reaches at least 10 degrees. According to Harald E. Esch of the University of Notre Dame, the winter cuculiinids can activate their central nervous system, and hence their shivering response, at unusually low temperatures. How they do so is still something of a mystery.

Heating up to 30 degrees C. from a resting temperature of zero degrees or below can drain a lot of precious energy. I therefore wondered whether the winter moths had a high metabolic rate, enabling them to generate heat more quickly and efficiently than other moths. They do not. As measured by oxygen consumption, their metabolic rates at rest and during shivering and flight were roughly the same as those found by other investigators for many moth species with a similar body mass.

The winter moths, in fact, pay a price in time and energy for heating themselves at low temperatures. When they warm up in an air temperature of zero degrees C., much of the heat they generate diffuses into the air, and in many instances they have to shiver for more than half an hour to reach a thoracic temperature of 30 degrees. During flight the large differential between the air temperature and the insects' body temperature results in accelerated heat loss, forcing the moths to repeatedly stop, shiver, and warm up again. In contrast, if the insects wait to shiver until the air temperature is nearly 10 degrees, they can warm up faster. Moreover, the heat they generate is sufficient to maintain nonstop flight at a thoracic temperature of from 30 to 35 degrees.

Lacking "low cost" ways of producing heat, the moths seem to be selective about the conditions under which they shiver. Intuition suggests that the animals should shiver when they feed, in order to be ready for a fast getaway from predators. Yet they sometimes do not do so. (Bats and birds may no longer patrol the skies in winter, but shrews, squirrels, and perhaps deer mice probably pose some danger.) Moreover, the insects do not heat themselves merely to keep warm; if they do not need to fly, they neither warm up nor resist cooling after a flight. In fact, the lower the air temperature is, the less likely the moths are to shiver. At from five to eight degrees C. only half (49 percent) of the moths in one study shivered as they lapped up diluted honey smeared on trees. In contrast, most of the moths (90 percent) shivered at approximately 17 degrees, the highest temperature at which I observed them in the field.

This apparent emphasis on energy conservation at the expense of predator avoidance makes sense if one examines the costs of resisting passive cooling. At an air temperature of zero degrees C. a moth weighing .1 gram (having a thorax weighing .04 gram) would have to counteract a postflight cooling rate of 13 degrees per minute to maintain a 30-degree difference between its body and the environ-

ment. In order to do so it would expend .42 calorie per minute.

If the moth in the example above filled itself to capacity on sugar-maple sap, obtaining four milligrams of sugar, its shivering would exhaust the contents of its stomach in just 35.2 minutes (each milligram of sugar provides 3.7 calories). At an air temperature of 15 degrees, on the other hand, the moth would burn calories much more slowly. In fact, it could maintain a 30-degree thoracic temperature for twice as long.

Because the moths do not have highly specialized mechanisms for producing extra warmth, I suspected that they must have an effective way of retaining heat. Actually they have several ways. In-

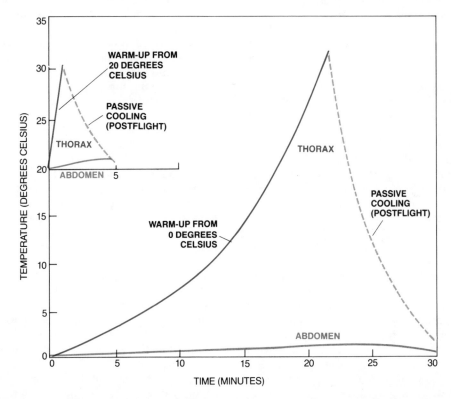

Figure 36 THORACIC AND ABDOMINAL TEMPERATURES of *Eupsilia* are plotted over time as the winter cuculiinid prepares itself for flight. When the air temperature is 20 degrees C. (*inset at left*), the moth can warm its thorax to 30 degrees within 1.5 minutes, whereas it must shiver for 22 minutes when the air is at zero degrees (*right*). In order to conserve energy, winter cuculiinids usually avoid heating up for flight when the air temperature is near zero. They also save energy by preventing heat from leaking into the abdomen (*blue*), which always remains within a few degrees of the air temperature. (The moths were held in place by a tether; they cooled rapidly when they were unable to take off.)

sulation can certainly retard heat loss, and the moths are well insulated by a coat of dense pile. (The pile is a derivative of the scales that give butterflies their beauty, and it explains why the moths are often called millers: the modified scales, which rub off easily, are whitish and fluffy, much like the flour that covers a miller.)

To determine exactly how well the scales facilitate heat retention, I measured the cooling rates of both pile-covered and depilated moths after they were heated and then exposed to varying air speeds in a wind tunnel. At air speeds of seven meters per second, which approximate those of flight, the unshorn moths cooled approximately twice as slowly as the naked ones.

Pile obviously helps the animals to store heat; indeed, it is absolutely essential to flight in the winter. Nevertheless, some summer moths of similar size, such as the tent-caterpillar moth *Malacasoma americanum*, have comparable insulation, to aid them when summer nights turn chilly. The insulation alone, then, does not explain why the winter cuculiinids withstand the cold better than other moths.

Like the insulation, an ability to prevent heat from diffusing out of the thorax into colder areas of the body would also help the moth to retain crucial thoracic warmth. Indeed, to varying degrees all endothermic insects examined to date have such an ability, including dragonflies, bumblebees, honey bees and many large moths. At low air temperatures, when they need to conserve thoracic heat, the animals retard heat flow to the head and the abdomen and virtually eliminate leakage to other extremities, such as the legs and wings. The winter cuculiinids do the same, but compared with other moths they lose even less heat to the abdomen.

Determining the thoracic and abdominal temperatures of cuculiinids required inserting fine thermocouple probes into them before they warmed up. During the preflight warm-up the abdomen remained within .4 degree C. of the air temperature. Indeed, the abdominal temperature increased an average of only two degrees even during flight, whereas the thoracic temperature increased as much as 35 degrees.

In collaboration with George R. Silver, formerly of the Cold Research Division of the U.S. Army Research Institute of Environmental Medicine in Natick, Mass., I also photographed the insects with an infrared camera. The camera records heat emission rather than external details. Our pictures confirmed that the legs, wings and abdomen of the winter moths gain little or no heat during warm-up, flight and postflight cooling.

How can a moth maintain a difference of more than 30 degrees C. between the thorax and the abdomen, trunk divisions separated by only one or two millimeters? Oddly, part of the answer lies in the anatomy of the moths' ears. The eardrums of noctuids lie behind the thorax, where they are enclosed by air chambers that happen to be almost perfect heat insulators. (No one knows whether the ears of winter moths still function as they once did to detect the ultrasounds emitted by bats.) Moreover, these chambers sit next to abdominal air sacs, which provide added insulation.

The vascular system also helps to prevent heat loss from the thorax. The blood, which transports nourishment stored in the abdomen, could potentially nullify the heat-sequestering ability of the air sacs. It flows in a single vessel from the abdomen (where the vessel is called the heart) through the thorax to the head. From the head it empties into the surrounding tissue, eventually percolating back to the abdomen.

In theory, blood returning to the abdomen could carry heat away from the thorax. In practice, one section of the circulatory system in the abdomen and one section in the thorax act as countercurrent heat exchangers: biological structures that in this instance recapture heat before it fully escapes from the thorax. In a countercurrent heat exchanger two fluids (or gases) contained within separate but adjacent channels flow in opposite directions. If the fluid in one channel is warmer than the fluid in the other channel, heat moves from the warmer substance into the cooler one.

The cuculiinid's abdominal heat exchanger lies under the moth's air sacs. It consists of both the blood vessel transporting cool blood from the abdomen to the thorax and a narrow region of the tissue surrounding the vessel. Blood that has been heated in the thorax flows through this tissue to the abdomen, which means that it moves in the opposite (or counter) direction from the vascular blood. Heat from the tissue therefore diffuses into the cooler blood that is flowing into the thorax.

As the blood vessel leaving the abdomen enters the thorax, the tube becomes the aorta—and the cuculiinids' second heat exchanger. Once inside the thorax, the vessel forms an inverted U, in which

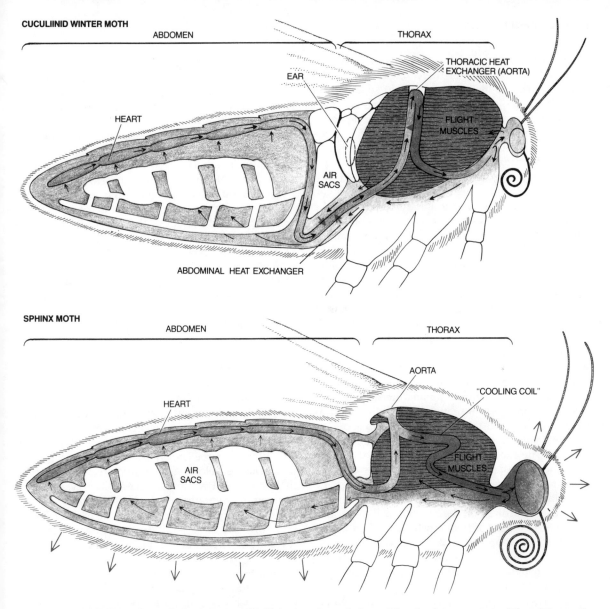

CUCULIINID WINTER MOTH

ABDOMEN   THORAX

THORACIC HEAT EXCHANGER (AORTA)

EAR

HEART

FLIGHT MUSCLES

AIR SACS

ABDOMINAL HEAT EXCHANGER

SPHINX MOTH

ABDOMEN   THORAX

AORTA

"COOLING COIL"

HEART

AIR SACS

FLIGHT MUSCLES

Figure 37 ANATOMY of a typical winter cuculiinid (*top*) differs from that of a summer moth (*bottom*) in ways that enable the cuculiinid to be active in the cold. The air sacs of the winter moth isolate the thorax, sequestering heat there. The circulatory system also preserves thoracic heat. In all moths the blood (*black arrows*) flows in a single vessel from the abdomen through the thorax to the head, warming along the way; on the return journey it percolates through tissue. The circulatory system of the winter moth includes an abdominal and a thoracic countercurrent heat exchanger. In the abdominal heat exchanger blood flowing between the heart and the aorta is cool (*blue*), whereas blood flowing in the opposite (counter) direction, through the adjacent tissue, is warmer (*red*); heat therefore passes (*red arrows*) from the tissue into the vessel, and then into the thorax. The thoracic heat exchanger is the aorta.

the two arms are pressed close together. First the vessel travels upward to the top of the thorax; then it curves sharply downward (before eventually turning away to the head). The blood flowing into the thorax from the abdomen is initially cooler than the thorax, but it becomes heated as it travels. The blood in the descending loop is therefore warmer than the blood in the ascending part. As a result heat returns to the ascending loop instead of traveling to the head with the flowing blood.

An ideal way to evaluate the effectiveness of a heat exchanger is to alter the configuration of the blood vessel. For instance, one might separate the ascending and descending parts of the thoracic heat exchanger and predict that heat from the descending part would then be lost to the head. Unfortunately it is next to impossible to do such surgery in a tiny moth without interfering with many things, including the rate of blood flow, which can in turn affect the movement of heat throughout the body.

It is possible, however, to judge the value of the winter cuculiinid's heat exchangers by comparing its vascular system with that of other moths, such as sphinx moths (Sphingidae) and giant silk moths (Saturniidae), both of which are large-bodied animals found particularly in the Tropics. In these insects the aorta is arranged to form a "cooling coil" instead of a heat exchanger. The descending part is greatly elongated and loops away from the ascending part. Rather than returning heat to the other part of the aorta, the looping vessel retains the heat and transports it away from the thorax.

The differing physiologies appear to produce markedly different effects. Sphinx moths and giant silk moths are up to 60 times as large (in mass) as the cuculiinids and therefore might be expected to overheat much more readily. On the contrary, they can dissipate excess heat to the head and the abdomen and from there to the air. In fact, they often fly at air temperatures higher than 30 degrees C. The cuculiinids, in contrast, never "dump" their excess warmth. Regardless of their very small size, they stop flying because of heat prostration when the air temperature approaches 20 degrees. Their extremely efficient heat-retention mechanism is apparently bought at a price, but one they seldom if ever have to pay.

Although the winter moths' heat exchangers effectively retain thoracic heat, the insects' circulatory system, like their pile covering, is not totally unique. Indeed, it is similar to that of many small-bodied summer moths. For instance, in the aorta of the tent-caterpillar moth, the descending loop is close to the ascending one, but the two are not pressed together. This small difference nonetheless appears to influence heat retention. The tent-caterpillar moth, which in flight maintains the same thoracic temperature as the winter cuculiinids, has a modest ability to dump heat. It can therefore fly in slightly warmer weather than the winter moths can, but it is not able to fly at low air temperatures.

Producing and retaining the heat needed for flight is just one part of the solution to winter survival. Because winter moths actually spend at least 99 percent of the time cooled down and in torpor, they also need some way to avoid freezing, or solidifying, when they rest in wait for a suitably "warm" night.

The immature stages of many summer insects survive the cold by producing biological antifreezes. I wondered if the winter moths did so too. By storing moths in a refrigerator for an average of three weeks, John G. Duman of the University of Notre Dame and I determined the standard freezing point of the insects' blood; that is, the freezing point when internal ice crystals are present. Even tiny crystals promote freezing because they provide a surface for the attachment of nearby water molecules.

The moths froze at from −1 to −2 degrees C., close to the freezing point of summer-adapted insects. The blood of moths that were freshly caught in the field had almost the same freezing point. These observations suggested that the cuculiinids produce little if any antifreeze; if they had such a substance in the blood, they would not freeze until the blood reached a much lower temperature.

It was still possible that the moths had a special ability to be supercooled (to remain unfrozen at temperatures below the standard freezing point) if they prevented an initial "seed" ice crystal from forming in or entering the body. The limit of supercooling is defined as the moment at which "flashing," or rapid ice-crystal growth, suddenly freezes the supercooled animal. This moment was readily identified by a quick, transient temperature rise in the moths; as water molecules join a spontaneously propagating ice crystal during flashing, they emit heat, which can cause a temperature rise of a few degrees. The flashing point is also the moment of death for the winter cuculiinids, because none survive freezing.

In moths that were chilled very slowly in an ice-free environment the supercooling limit varied markedly, ranging from −4 to −22 degrees C. Moreover, the moment of freezing did not cluster around any one temperature. Such variation suggests that supercooling is a random phenomenon, not an adaptive trait, in the winter moths. Indeed, the supercooling ability of the winter moths is similar to that of many summer moths that never encounter freezing temperatures.

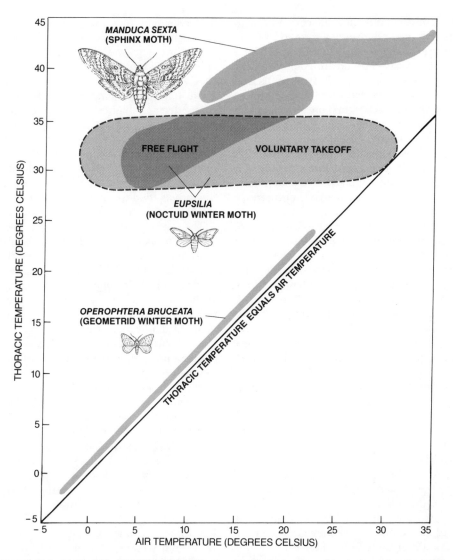

Figure 38 THORACIC AND AIR TEMPERATURES at which summer and winter moths engage in continuous flight. *Manduca sexta,* a summer sphinx moth, must heat its thorax to about 40 degrees C. before flight; it can remain aloft in air temperatures as warm as 35 degrees C. It cannot initiate flight until the air reaches about 12 degrees. *Eupsilia* can take off in the range of −2 to about 30 degrees (*gray*), but when the air is cooler than five degrees, the moth stops flight repeatedly to shiver. When the air is warmer than about 20 degrees, it overheats. *Operophtera bruceata* can fly in air temperatures ranging from −3 to 25 degrees and does not have to heat its thorax much. It overheats at air (and thoracic) temperatures close to 25 degrees.

Although supercooling is probably not a specific adaptation for winter survival, any ability to cool below the standard freezing point may enable insects to survive low air temperatures, provided they stay in a dry place away from ice that might enter the body and serve as a seed crystal. A small, dry cavity in the ground, in a rotting log or under bark often forms an ideal habitat for hibernating insects. Where do the winter moths go to escape ice and sharp declines in their temperature?

An observer who knew that coloration often provides protection might well suppose the moths rest on trees. Why else would they, in common with their summer-adapted relatives, be colored in ways that provide camouflage against tree trunks? There are, for instance, white- and cream-colored moths that are nearly invisible on birch; brown moths that blend with fir or spruce; black varieties that almost disappear against ash bark; gray ones that match beech or elm, and even some moths that are peppered to resemble lichens.

I discovered the moths' resting place by first building a large outdoor enclosure and standing in it the trunks of pine, spruce, birch, beech, ash, maple, elm and cherry; a leafy layer covered the ground. I then released 173 moths of varying colors one evening and searched for them the next morning. Most of the recaptured moths had crawled under the leaf litter or into curled leaves lying on the ground.

Leaves provide excellent insulation from the cold. In late winter in Vermont I measured temperatures of no less than −2 degrees C. under the ground cover even when the air temperatures dipped below −30 degrees. Such a makeshift shelter is often covered with snow, which provides an added but perhaps unnecessary barrier against the cold. Dale F. Schweitzer of Yale University has shown that the fallen leaves can be insulation enough, at least at temperatures as low as −23 degrees.

Moths that hide under leaves should not need camouflage. The explanation for their coloring may be found in their evolutionary history. The fact that the external and physiological features of the winter moths are quite similar to those of summer moths suggests that the winter-adapted variants evolved from ancestors that were active in the summer. If that is the case, the winter moths probably did perch on trees at one time. After they adopted new habits their color became a neutral trait and was not altered. Assuming that this hypothesis is correct, it suggests that a "reversed" life cycle evolved independently many times. Indeed, according to John G. Franclemont of Cornell University, taxonomic findings also suggest that the winter cuculiinids are polyphyletic (descendants of more than one ancestral line) and have all developed the same strategy for winter survival.

The moths frequently seek shelter to avoid freezing when they rest, but at other times they may optimize their energy balance by avoiding too cozy a hiding place. Indeed, the lower they keep their body temperature while resting (short of freezing), the longer they can make their energy reserves last.

The reason is that the metabolism slows in cold weather. For example, on the basis of measurements of energy metabolism in resting moths I calculate that a moth weighing .1 gram and filled with six grams of sugar from sap can rest for 193 days at an air (and body) temperature of −3 degrees C. At three degrees higher, or zero degrees, the fuel would last for only 24 days, and at 10 degrees the reserve would be exhausted in just 11 days. Whether or not the moths actually attempt to remain at the lowest temperature possible is not yet known. To do so would entail an important risk: if the insects tried to sit still in an exposed place, they might become too cold and freeze.

The accumulated evidence suggests that the moths do not have any highly specialized adaptations for resisting the cold, just as they have no unique adaptations for generating heat. On the other hand, their behavioral adaptation of seeking shelter in the leaves serves them nicely. It allows them to be flexible. They can become active the moment the air is warm enough for efficient flight, and yet they can also find shelter within seconds if temperatures plummet dangerously during the same evening. Insects with adaptations enabling them to survive freezing by perfusing themselves with antifreezes would require a significantly longer lead time to become fully active—time that winter evenings may not provide. In high concentrations antifreezes (primarily alcohols) are toxic and temporarily leave animals comatose; the substances are eventually converted into less toxic chemicals, but that happens slowly, particularly when the temperature of an animal is very low.

Winter cuculiinids seem to be well adapted to cold weather, but it would be a mistake to conclude that their characteristics are the only ones possible for winter activity. On frosty nights in New England one can see the males of the species *Operophtera bruceata* sailing through the forest in search of the flightless, sluglike females. The moths, which also fly on sunny days, are even active at temperatures as low as −3 degrees C. and during mild snowstorms in November. (In late November, before they disappear entirely, the males fly only at about noon on sunny days.)

Operophtera, which is one of a few species in the Geometridae family that have adapted to winter, neither basks (a typical warming behavior of day-flying insects) nor shivers, nor does it have the insu-

Figure 39 MALE *OPEROPHTERA* (*left*) has large wings, a feature that helps to explain why it can fly without significantly warming its wing muscles. In contrast to the winter cuculiinids, which beat their wings approximately 60 times per second, male *Operophtera* moths beat their wings as few as two times per second. The female of the species (*right*) does not fly.

lation found in cuculiinids. Instead the males are able to function at an extremely low body temperature; they are the only moths that routinely fly with a muscle temperature close to zero. Spared from the need to warm up before takeoff, *Operophtera* saves the considerable energy that might otherwise go toward shivering.

Oversized wings and a low body weight contribute to such energy conservation by enabling the insects to remain airborne with a wingbeat frequency as low as from two to four beats per second —much lower than the more than 60 beats per second required by the cuculiinids. In common with many winter-adapted insects, the adults do not eat; in fact, they no longer have a digestive tract. (All the energy they use is accumulated and stored at the larval stage.) It is impossible to ascertain cause and effect, but the loss of the need to carry a digestive tract probably reduced the energy needed for flight.

Geometrids that live near equatorial lowlands have physical characteristics similar to those of *Operophtera*, but they are less exaggerated. As is true of the cuculiinid winter moths, *Operophtera's* design for cold-weather activity appears to be modeled on a preexisting form, albeit one quite different from that of the noctuids.

The remarkable ability of certain noctuids and geometrids to be active in the winter highlights the way slight evolutionary alterations in anatomy, physiology and behavior can add up to success in a new environment. The winter moths are still quite similar to their close relatives but, taken together, their small differences adapt the insects to winter living.

The Physiology of the Koala

This gentle marsupial eats mainly eucalyptus leaves (which are toxic to other animals), seldom drinks and uses no shelter. Recent studies indicate how the animal manages to survive in its distinctive niche.

• • •

Robert Degabriele

July, 1980

A common reaction of people who see a koala (or more likely a picture of one, since the only place outside of Australia where koalas live is the San Diego Zoo) is to perceive the animal as a living teddy bear and to wonder if they could have one as a pet. The koala is of course not a bear and they cannot have one. It is a marsupial, a primitive mammal whose young are born in a virtually embryonic state and spend the early part of their life in a pouch that covers the mother's teats. And although it was once possible to keep koalas as pets, the practice is now forbidden by rigorous protective laws enacted in Australia because the animal was hunted for its pelt.

The scientific name of the koala is *Phascolarctos cinereus,* from the Greek for pouched bear and ash gray. The animal is formally characterized as an arboreal folivore, meaning that it lives in trees and eats leaves. The koala's tree is the eucalyptus, specifically 35 or more of the 600 or so species of the genus *Eucalyptus* that grow in Australia. The diet of an adult koala is almost exclusively eucalyptus leaves, and since the oils of eucalyptus leaves are toxic to most other mammals, one wonders what adaptations enable the koala to thrive on them. This subject has been the focus of my own work with koalas, and I shall be returning to it.

A fully grown koala weighs on the average about nine kilograms (20 pounds) and its body is at most about 62 centimeters (two feet) long. Its fur is thick and woolly, its limbs are long and its toes are strongly clawed. On each front foot the two innermost digits can be opposed to the others like two thumbs, as can the innermost digits on each rear foot. It is this articulation that makes the koala a good climber. On the ground, to which it normally descends only to move from one tree to another, the koala goes on all fours. Most of the time it is in a tree, sleeping and foraging among the leaves intermittently both by day and at night.

Koalas usually breed every second year. The mating season extends from early spring to midsummer, and the gestation period is about 35 days. A newborn koala (usually a single birth) is about 19 millimeters (three-quarters of an inch) long and weighs about 5.5 grams (less than a quarter of an ounce). This tiny creature climbs unaided to the pouch, which is unusual in that it opens to the rear, and fastens onto a teat. About six months later, having attained a length of some 20 centimeters and grown a good coat of fur, the young koala emerges from the pouch and thereafter, for another six months, is carried on its mother's back. At six months of age it is weaned on the emulsified droppings of the mother and then is ready for its adult diet of eucalyptus leaves.

For mammals a diet of leaves is a poorer source of energy than many other kinds of food would be. Moreover, mammals are less efficient than other animals in digesting leaves. For this reason many mammalian arboreal folivores are thought to be living close to the limits of their energy budget. Therefore even though the koala is biologically successful, the foundation of its success is precarious.

The question of the role of the volatile oils of eucalyptus leaves in the feeding behavior of the koala has taken up much time and effort. The common observation that the preference of the animal varies from species to species of eucalyptus and even from tree to tree within a species has led to the widely held assumption that a koala actively chooses leaves on the basis of their oil content and has a particular reason for doing so. The reasons put forward for the choice include the antimicrobial action of the oils in the voluminous cecum of the koala's intestinal tract and the "need" for the heat-generating effect of the oils in maintaining body temperature.

Ian Southwell of the Museum of Applied Arts and Sciences in Sydney recently examined the relation between the koala and the composition of volatile oils in eucalyptus leaves. He found no correlation between variations in the level of the volatile oils in the leaves and the feeding preferences of koalas. What his finding demonstrates is that the koala's considerable success in occupying the eucalyptus habitat is due to a general ability to detoxify oils that are harmful to other animals. In other words, the koala is one marsupial that has been able to overcome the chemical defense mechanism of eucalyptus trees.

In this way the koala has gained access to an abundant resource. How does the animal utilize it? How adequately does a eucalyptus provide food, water and shelter?

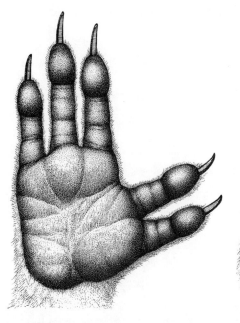
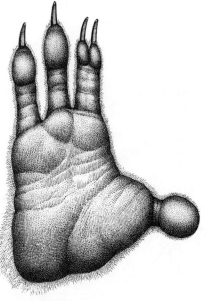

Figure 40 HANDS AND FEET of the koala are adapted for movement in trees by having long and strong claws and digits that are opposable to other digits as the human thumb is. The hand (*left*) has two such digits, the foot one digit. The joined second and third toes of the foot are often employed by the animal in combing its fur. The palms and soles have granular pads for gripping.

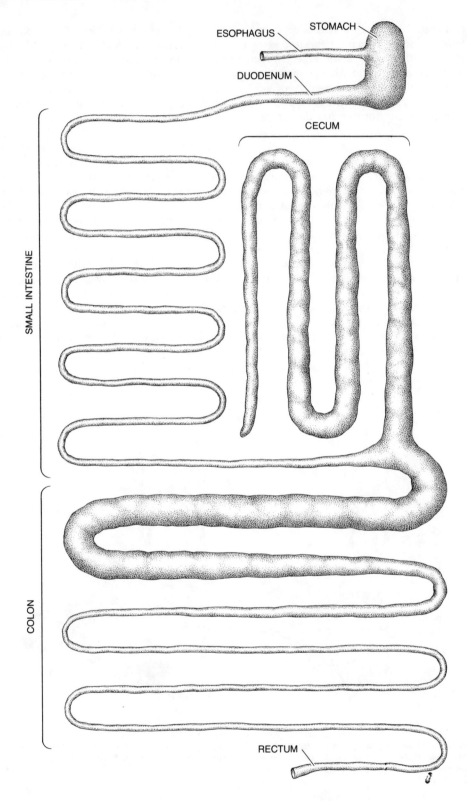

The first step in answering these questions is to consider the anatomy and physiology of the koala's digestive system. In common with other herbivorous mammals the koala is unable to digest cellulose and so must rely on microorganisms that do it. The location of the microorganisms in the digestive tract provides a basis for classifying herbivorous mammals into pregastric and postgastric digesters. In the pregastric category are such eutherian (placental) mammals as cattle and such marsupials as the kangaroo and the wallaby; the postgastric group includes such eutherian mammals as the horse and the rabbit and such marsupials as the brushtail possum and the koala.

Among the postgastric digesters the commonest site for the microorganisms is the cecum, an expansion of the hindgut at the junction of the small and the large intestine. The cecum is the most remarkable feature of the koala's digestive system. Recent observations have established that it takes up about 20 percent of the length of the postgastric intestine. Thus the koala has a fermentation chamber where the passage of leaves can be delayed so that the microbes can digest the cellulose.

The possession of such a large cecum means that the koala's dietary requirement for carbohydrate is likely to be met at all times through the microbial digestion of cellulose. The animal's position with respect to nitrogen, that is, protein, is less assured. In order to determine the koala's ability to remain in positive nitrogen balance my colleagues and I at the University of New South Wales in Australia maintained a group of koalas in cages, feeding them exclusively on fresh leaves of the gray gum (*Eucalyptus punctata*), which is known to be a food tree for koalas. We measured the nitrogen taken in and excreted in both summer and winter.

As one might expect, we found that in such a regimen the koala is quite capable of remaining in positive nitrogen balance throughout the year. The way the balance is achieved was not so predictable. Although the intake of digestible nitrogen seemed

Figure 41 ALIMENTARY TRACT of the koala is notable for the size of the cecum, which accounts for some 20 percent of the length of the intestine. Since the koala cannot digest the cellulose in the leaves that are its staple diet and relies on microorganisms to do it, the cecum serves as a fermentation chamber in which the passage of leaves through the digestive tract can be delayed so that the microbes can do their work. The koala can detoxify the oils of eucalyptus leaves.

to be about the same at all times, the dietary intake was significantly higher in winter. In other words, the koalas had to consume more leaves in winter to remain in nitrogen balance.

The explanation probably lies in seasonal differences in the quality of the leaves. Eucalyptuses such as the gray gum grow rapidly in spring and early summer but not much after they flower. It is therefore likely that winter leaves are older and more fibrous and contain less digestible nitrogen. Here is one bit of evidence that mammalian arboreal folivores may be living close to the limits of their energy budget.

The precarious nature of the koala's energy budget is further emphasized by the phenomenon known as "wasting disease." In times of drought koalas have become comatose and have died with a full stomach. I suspect that a nitrogen deficiency is the major factor, since during a drought eucalyptuses grow few leaves or none. The remaining leaves are comparatively old and contain progressively less digestible nitrogen. A koala's response to such a change in the quality of the available leaves is likely to be to eat more leaves. If the quality of the leaves deteriorates enough, the quantity of leaves that is required and even the capacity of the koala's digestive system can become limiting factors. Under such circumstances it may become physically impossible for a koala to satisfy its nutritional requirements.

The koala also has a reputation for seldom or never drinking water. This is a common theme running through the various legends of the Australian aborigines about the koala. Some of the aboriginal names for the animal, such as *koobor*, mean "Does not drink water." Curiously, the name koala, an aboriginal name from the Hawkesbury River district near Sydney, has no such connotation.

The available evidence suggests that in normal conditions a koala gets the water it needs from dew and from eucalyptus leaves. In general water is lost in the urine, in the feces (either as free water or as water that would have been produced by the metabolic breakdown of food) and through evaporation. The urinary water loss is regulated by the kidney. The free-water content of the feces is regulated by the large intestine, and the metabolic-water content depends on digestive efficiency. The evaporative loss is tied closely to the animal's mechanisms for regulating its body temperature.

However the koala's water balance is achieved, the turnover rate of water is useful as an indication of both the availability and the utilization of water. One can determine that rate (and also the total content of water in the body) by observing the dilution and the rate of disappearance of water labeled with tritium, the radioactive isotope of hydrogen. We studied the water metabolism of free-ranging koalas in three widely separated populations: one at Magnetic Island, near the northern limits of the koala's range, one at Sydney, near the center of the range, and one at Phillip Island, the southern limit of the range. No significant differences were found in total body water and turnover rate.

The finding indicates that the microhabitat of the koala is reasonably uniform in terms of water requirements and water supply no matter where the animals live. The free-water content of gray-gum leaves ranges from 40 percent (old and fibrous leaves) to 65 percent (new growth). The leaves of other species of eucalyptus contain at least as much water as those of the gray gum, since leaves with a water content below 40 percent dry out and die. One can conclude that in normal conditions the leaves of eucalyptus trees provide the koala with both food and water in adequate amounts.

The koala's content of body water is relatively high (77.4 percent of the animal's weight). That is close to the water content of the fat-free component of the carcass of any mammal, a relation suggesting that the total absence of fat deposits in all the koala carcasses I have examined is characteristic of the koala. The lack of fat may be a consequence of the koala's precarious nutritional balance. It is also what enables the animal to carry such a high proportion of body water. (Much of the water is in the cecum, which can hold large amounts of moist food.)

The usefulness of carrying a considerable amount of water becomes clear when the rate of water turnover by the koala is examined closely. When that rate is expressed in terms of the four-fifths power of body weight, the effect of varying body size is negated and one can make comparisons within and between species. We have compared the koala and

Figure 42 BROWSING KOALA is seen in its characteristic habitat, a eucalyptus tree. Koalas subsist on the leaves of about 35 of the approximately 600 species of the genus *Eucalyptus* that grow in Australia. The leaves provide all the nutrients the koala needs and usually enough moisture so that the animal does not have to drink. The koala feeds and sleeps intermittently in the trees throughout the day and night.

the short-nosed bandicoot (*Isoodon macrourus*), a rat-size terrestrail marsupial that feeds on invertebrates. The water-turnover rate of short-nosed bandicoots living on an island with no available drinking water is similar to that of the koala (179 grams per .8 kilogram of body weight per day). In bandicoots living on the water-rich mainland the rate is much higher (243.8 grams). This comparison reinforces the idea that the koala's food incorporates a supply of water and that in normal circumstances the animal does not drink water. The idea is further reinforced by the fact that the koala has a simple type of kidney that is not capable of a high degree of water conservation. Animals that have evolved to survive with a low intake of water typically have a kidney that can produce highly concentrated urine, so that more of the available water stays in the body.

We investigated the way the koala achieves water balance by making measurements on individually caged koalas. Each koala was given a daily supply of freshly cut branches of gray gum. Drinking water was always available in the summer but was sometimes withheld in the winter. Inputs and outputs of water were determined as changes in weight.

The rate of water turnover of the caged koalas was no more than half the rate of their free-living counterparts. The most likely reason for the difference is the reduction in activity imposed by a cage. As would be expected in an arboreal mammal, the leaves were the major source of water.

Although the koalas consumed more food and water in the winter, drinking water contributed only about a fourth of the intake in both summer and winter. It is likely, then, that the increased intake of water in the winter is the result of a need to consume a greater volume of nutritionally inferior leaves. In other words, nutritional requirements would dictate the level of water intake. Nevertheless, the water intake associated with the food remains reasonably constant at from 40 to 50 grams per kilogram of body weight per day. Therefore, provided that the leaves being eaten are nutritionally adequate, the koala can get enough water from its diet.

The koala loses water mainly through evaporation from its respiratory surfaces. The loss of water in the urine is the least significant component. Evaporative loss and the production of urine show remarkably little variation between seasons and when drinking water is not available. The slight decrease in urinary output in the absence of drinking water suggests that the koala has a water-conservation mechanism of the kind found in the pig and the beaver, whose kidney actively reabsorbs urea when drinking water is limited. In the koala, however, the reduction in the volume of urine is so slight that I suspect the urea is being reabsorbed for nutritional reasons rather than for water-conservation ones.

The loss of water in the feces is probably the most effective means the koala has of regulating its water balance. The feces are always quite dry. When drinking water is not available, the water content of the feces drops from 52 to 43 percent, which is the same as the water content of the feces of a dehydrated camel. It is evident that the koala can manage quite well with a regularly available but not overabundant water supply.

The high rate at which water is lost through evaporation from the respiratory surfaces is an indication of the importance of the relation between water balance and thermal balance. An examination of the way in which the koala regulates its body temperature clarifies this relation.

The koala is unusual among the arboreal marsupials in that it does not seek any kind of shelter, a pattern of behavior (or rather a lack of one) found in only one other arboreal marsupial group, the tree kangaroos. (Such behavior is quite common among tree-dwelling placental mammals.) One is therefore stimulated to find out what kind of protection against extremes of environmental conditions is provided by the fur of the koala.

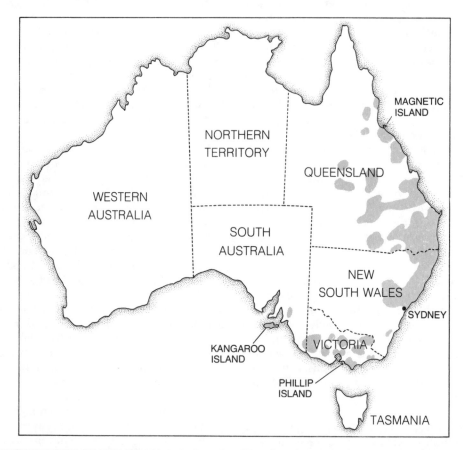

Figure 43 DISTRIBUTION OF THE KOALA is indicated by the colored areas on this map. The animal was once so vigorously hunted for its pelt that it was threatened with extinction. With the enactment of protective laws Australia also made efforts to reestablish the species, particularly in Victoria, where colonies were set up on islands and in mainland forests. The northernmost colony appears to be one that developed from a group of koalas introduced to Magnetic Island.

The dorsal fur, which is the thickest at 54.4 hairs per square millimeter, covers 77 percent of the animal's body surface. The ventral hair has only half the density of the back fur and covers 13 percent of the body surface. The differences in density are not paralleled by differences in hair length, which for both the longer guard hairs and the shorter underfur is virtually the same all over the body. One does, however, find seasonal differences in hair length: the length of the long hairs differs from that of the short ones more during the summer than during the winter.

The thick dorsal fur is darker than the sparse ventral hair; it therefore tends to absorb solar heat as well as to provide insulation. The sparse ventral hairs can be erected, so that the amount of insulation the ventral hair provides can be adjusted. The combination of these kinds of covering gives the koala a type of environmental control reminiscent of that of the guanaco, a llama-like South American animal with a densely matted dorsal surface (40 percent of the total area) and sharply defined areas of almost bare skin on the ventral surface. By changing its posture the guanaco can achieve as much as a fivefold variation in thermal insulation in still air and five times more than that in the wind.

Casual observations of koalas in trees on windy days indicate that as the wind speed rises a koala tends increasingly to roll up into a compact ball in order to present only its curved middorsal surface to the wind. Koalas did this even when the temperature was high. As the wind speed increased further their ears were folded forward so that almost nothing projected into the airstream.

The picture of a portable shelter worn by every koala was enhanced by measurements we made of the insulation characteristics of koala pelts. (Our sample was small and, being fortuitous, consisted of more summer pelts than winter ones, but there was not much difference between the two kinds.) The koala's dorsal fur proved to have the highest insulation value among the 12 marsupials so far investigated. Indeed, it is in the lower range of the values obtained for arctic animals.

The effect of wind on the insulating power of the dense, matted dorsal fur is small. Moreover, compared with the fur of a number of other animals the koala's fur shows the smallest decrease in insulating value as the wind speed increases, at least up to about 10 miles per hour. The average decrease was 14 percent and the minimum was a remarkable 3 percent. These data suggest both that the fur would maintain a significant level of insulation at much higher wind speeds and that for an animal living in the exposed treetops of the open forest the fur provides excellent thermal protection.

The temperature-regulating effect of fur is complemented by metabolic activity. The koala's basal metabolic rate is 74 percent of the rate predicted for marsupials in general. (Among placental mammals such folivores as the sloth and the potto show a similar divergence from the predicted rate.) The koala responds to high environmental temperatures by panting. At low environmental temperatures the low metabolic rate is accompanied by a high level of total body insulation, with the fur contributing 50 percent (a rather large proportion). The metabolism-insulation pattern of the koala appears to be characteristic of many tropical arboreal mammals; perhaps the arboreal marsupials in general differ from the terrestrial ones in this way.

In examining the relation between water balance and thermal balance I began with the fur and led into thermoregulatory mechanisms. Now I can complete the circle by considering the relation between evaporative water loss (a thermoregulatory characteristic) and metabolic water production (an aspect of water balance). The amount of water produced by metabolism for every gram of oxygen consumed is calculated on the basis of the nutrient composition of the leaves and of the feces. The ratio of metabolic water production to evaporative water loss is then derived from the relation between oxygen consumption and each of those processes. Such calculations have shown that the koala's requirements for evaporative cooling are adequately provided for up to an ambient temperature of about 30 degrees Celsius (86 degrees Fahrenheit), which is rarely exceeded for long in the animal's environment.

In summary, the koala is an animal whose ecological niche can be described (in the most general sense) as the arboreal environment provided by trees that belong mainly to the genus *Eucalyptus.* They provide the animal with a supply of food and water and with a place to live. With its specialized digestive system the koala can overcome the toxic effects of eucalyptus oils and can extract enough nutrients and water from the eucalyptus leaves. The koala's thermoregulatory pattern is attuned to that level of water supply, and so (other considerations aside) the animal is potentially able to thrive in most of the forests of Australia.

Carnivorous Plants

The luring, capturing and digesting mechanisms that have evolved in some plants enable them to devour insects in order to augment their supply of mineral nutrients and thus survive in habitats where few other plants can live.

. . .

Yolande Heslop-Harrison

February, 1978

Green plants gain their energy from the sun, their carbon from the atmosphere and their water and mineral elements from the soil. The atmospheric carbon (in the form of carbon dioxide) and the soil nutrients are replenished from the wastes of microorganisms and animals, and in this way plants and animals are complementary in the general economy of nature. A few plants, however, have evolved the capacity of feeding directly on animals, supplementing their nutrition by capturing and digesting animal prey. By adopting this habit they have gained the ability to survive in nutrient-poor environments, although in some instances at the expense of being unable to exist in richer habitats in competition with species that have a more usual life-style.

Those flowering plants that have evolved the carnivorous habit can be divided into two groups according to their methods of catching prey: active trappers and passive trappers. Of the active trappers *Dionaea muscipula*, the Venus's-flytrap, is one of the most familiar. In nature this species is found only in certain habitats on the coastal plains of North and South Carolina. Now, however, it is also widely cultivated and can even be seen on sale as a novelty at the checkout counters of supermarkets. Its natural prey are mainly hopping or crawling insects and spiders. Prey touching the leaf agitate tactile hairs; the action triggers a closing mechanism and the hinged leaf snaps shut.

A closely similar mechanism of trapping, although on a diminutive scale, is found in another genus of the same plant family: *Aldrovanda*, the waterwheel plant. As its common name indicates, *Aldrovanda* is an aquatic plant. The genus includes

Figure 44 ACTIVE AND PASSIVE TRAPPERS among carnivorous plants. First is the bladderwort. When swimming prey touch the trigger hairs around the mouth, the flap of tissue that forms the door swings open and the bladder expands suddenly to draw in both water and prey. Second is the Venus's-flytrap. When a visiting insect or spider touches one of the trigger hairs on the leaf surface, the two sides of the leaf move quickly together, closing the trap. Third is a passive trapper, the pitcher plant. The prey is lured to the pitcher's slippery edge, falls into a pool of digestive fluid and cannot climb out. Last is a sundew. Small flying insects that alight on the attractively colored leaf surface touch the secretion globules, which secrete the digestive enzymes.

only one species, *A. vesiculosa*, but it has a wide-spread distribution, occurring in central and southern Europe and eastward across Asia into Japan and parts of India and Australia. The commonest of all the active trappers belong to the genus *Utricularia*, which includes some 150 species. In the aquatic or semiaquatic species of this genus the traps take the form of small elastic-walled bladders, hence the common name bladderwort. When the bladder is "set," it is flattened, and the entrance to it is sealed by a flap of cells. The prey are swept into the bladder with the current of water produced when the walls spring apart following the opening of the entrance flap, an action that is triggered by tactile hairs near the entrance.

The passive trappers include the pitcher plants, where the prey is captured and digested in pitcher-like structures formed by modification of the entire leaf, as in the North American genus *Sarracenia*, or by an extension of the leaf tip, as in the genus *Nepenthes* of the eastern Tropics. The prey are enticed to enter the pitcher by colors and scents, much in the way that pollinating insects are attracted to flowers, and they are then drowned and digested in the pitcher fluid. A different strategy is seen in plants with flypaperlike leaves, as in the species of *Pinguicula*, the butterworts, and *Drosera*, the sundews. In these genera glands on the leaf surface secrete adhesive droplets. The prey, usually a flying insect attracted by odor or color or perhaps by the brilliant refractions of the droplets, is trapped by the adhesive when it alights. It becomes more firmly attached to the leaf surface as its efforts to escape bring it into contact with more glands.

The effectiveness of carnivorous plants as predators has been well documented for several species, and the published lists of animal species captured are remarkably long. The prey are usually quite small, but mice have been found in the pitchers of *Nepenthes*, probably the victims of a chance fall, and the remains of small tree frogs have been found in the pitchers of *Sarracenia*. Bladderworts sometimes catch small fishes and tadpoles, but the bladders are never more than a few millimeters wide and are more adapted to the capture of rotifers, copepods, and the aquatic larvae of such insects as mosquitoes.

The quantity of prey captured is sometimes quite large. In the pitcher plants the pitchers usually survive for several months, and they may be virtually filled with the decaying remains of their catch. In plants with more ephemeral traps, such as the species of butterworts in which the effective life of a leaf may be only five days, the total catch of a growing season is more difficult to assess. A butterwort such as *Pinguicula grandiflora* grows one new leaf about every five days, so that a total of 400 square centimeters of catching surface may be produced in a single season, even though the diameter of the leaf rosette never exceeds eight centimeters. Carnivorous plants also occasionally form dense stands. Some 30 years ago Francis W. Oliver of University College London described a sward of sundews extending over an area of more than two acres near Barton Broad (not far from the Norfolk coast in eastern England) that had captured vast numbers of butterflies, most of them cabbage whites trapped when the settled after a migratory flight from the Continent. Oliver found four to seven butterflies adhering to each plant and estimated the total number of trapped insects to be about six million.

What, then, is the value of the carnivorous habit in plants? Charles Darwin, who was one of the pioneers of work on the physiology of carnivorous plants, addressed himself to the question a little more than a century ago. With his son Francis he showed convincingly that sundews in cultivation that had been fed artificially by applying insects to their leaves were more vigorous, produced more flowers and set more seed than those denied such fare. More recently Richard Harder of the University of Göttingen and others have shown that butterworts, sundews and bladderworts cultivated in controlled environments with carefully regulated access to nutrients perform better when provided with prey, confirming the Darwins' results. They have also found that butterworts make use of pollen carried in the atmosphere, digesting it much as they digest insects.

The nutrients derived from captured prey enter the leaves with surprising rapidity. Some years ago Bruce Knox and I used algal protein labeled with the radioactive isotope carbon 14 in order to follow the movements of the products of digestion in the plant. The leaves of butterworts were fed minute quantities of the labeled protein, and the breakdown products were traced by means of autoradiographs. We found that the amino acid and peptide products of protein digestion moved into the leaf in two or three hours and then passed into the stem and on toward the roots and growing points in less than 12 hours.

The main pathway through the leaf was the xylem, the water-conducting tissue of the plant.

Recently John S. Pate and Kingsley Dixon of the University of Western Australia have labeled fruit flies by feeding them on yeast containing the isotope nitrogen 15 and have used the flies as a diet for *Drosera*. In the sundews growth arises from corms formed during the preceding season, and much of the nitrogen reserve in the corms is in the form of the amino acid arginine. Pate and Dixon found that at the end of their experiment some 40 percent of the arginine in the corms of the experimental plants contained nitrogen 15, a striking demonstration of how important the supplemental nutrients obtained from the prey must be for the survival of the species in nature.

About a quarter of a million species of flowering plants exist on the earth today, and of them some 400 are known to be carnivorous. They belong to 13 or so genera of six families. Some of the families are quite diverse, with members on every continent. If the main function of the carnivorous habit is to provide scarce nutrients, one might expect the plants to inhabit the kinds of environment where such supplementation would be most beneficial. That is just what is found. The plants are most often encountered in nutrient-poor communities: on heaths or in bogs, on impoverished soil in forest openings and occasionally on marl, the crumbly clay soil associated with weathered limestone. Often two or three different genera of carnivorous plants will be found growing together in such localities. For example, in the Pine Barrens of New Jersey several species of sundew, pitcher plant and bladderwort coexist.

At the same time some carnivorous species occupy remarkably narrow ecological niches. Certain bladderworts of South America are found only in the pools of water that accumulate in the natural basins formed by the leaf rosettes of bromeliads, members of the pineapple family. In this environment they are essentially free from competition. Species of the pitcher-plant genus *Heliamphora* provide another example. They occur in nature only in the remote mist zone high in the mountains along the border between Guyana and Brazil, and their environmental requirements are so peculiar that the plants can be maintained in greenhouse cultivation

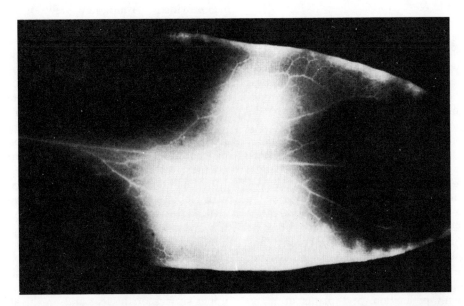

Figure 45 MOVEMENT OF DIGESTED MATTER from a leaf surface back into the plant is traced in this autoradiograph of a butterwort leaf. A small amount of protein labeled with the radioactive isotope carbon 14 was placed on the surface of the leaf at one side of the midrib. After eight hours the radioactive digestion products (amino acids and peptides) had spread over much of the leaf surface (*large bright areas*). Movement of products in vascular system of leaf, mainly toward stem but also toward leaf margin and tip, is indicated by thin bright lines.

only with difficulty. Still another example is the cobra plant of North America, the only species in the pitcher-plant genus *Darlingtonia;* its distribution is limited to mountain slopes and coastal bogs in Oregon and California.

A number of the carnivorous plants are able to survive in quite extreme environments. *Sarracenia purpurea,* one of the nine North American species of this pitcher-plant genus, is found from Florida northward to British Columbia and Newfoundland. At the northern limit of its range the bogs where it grows may be frozen for several months each winter. In Australia (where more sundew species are found than are found anywhere else in the world) some of the sundews native to the northwest are exposed to near-freezing night temperatures during the wet winter months when they are actively growing and become dormant during the summer dry season, when the daytime temperatures in the shallow depressions in granitic rocks where they live may exceed 50 degrees Celsius (120 degrees Fahrenheit). The dewy pine of the western Mediterranean, a passive trapper and the only species in the genus *Drosophyllum,* is relatively drought-resistant during its main growing period. Found on the dry, sandy coastal plains of Portugal and Morocco, it depends on sea mists for a part of its water supply.

Contemplating the range of adaptations found together in the plant carnivores, one can readily appreciate Darwin's absorption with them as examples of evolutionary virtuosity. The trapping mechanisms themselves represent elaborately modified leaves or leaf-like organs, and they are usually associated with lures and guides that tempt and direct the prey into or onto the trap. Specialized glands secrete the digestive enzymes, and the same glands or others retrieve the products and pass them back into the plant for distribution through the conducting tissues to the sites of growth. None of the individual features—the traps, lures, odors, directional guides, secreting glands, and absorbing glands—is itself peculiar to the carnivores. Many plants have leaf parts capable of rapid movement, for example *Mimosa pudica,* the sensitive plant; others have elaborate insect-trapping mechanisms associated with pollination, and plants of many families have glands capable of secreting water, salt, mucilage, sugars, proteins and other products. It is the assemblage of features that gives carnivorous plants their unique character, the bringing together of so many individual adaptations into a functional combination directed to an end so unusual for a photosynthetic plant that it seems grotesque and even macabre.

Students of plant morphology have long been interested in the special features of the carnivorous plants. Notable among them in Britain, besides Darwin, was Joseph Dalton Hooker, at that time director of the Royal Botanic Gardens at Kew, and his assistant, William Thiselton-Dyer, who succeeded Hooker as director. In Germany they included Karl I. E. von Goebel, a leading plant morphologist and anatomist, and C. A. Fenner. In North America, Francis E. Lloyd later contributed many detailed observations, notably on the trap of the bladderworts and its firing mechanism. Lloyd published his classic study, *The Carnivorous Plants,* in 1942; it is still a standard in the field today.

Little could be added to the remarkably precise work of these earlier observers until the advent of electron microscopy. The transmission electron microscope has revealed many features of subcellular structure connected with the processes of secretion and absorption in carnivorous plants, and in the past decade the scanning electron microscope has provided revealing new views of the traps and their associated glands.

The traps of the different types of carnivorous plants have several kinds of surface glands, some concerned with capture and digestion and some with other functions. Certain glands produce nectar as a lure for prey, much as the nectar-secreting glands in flowers that attract insect pollinators do. In *Nepenthes* such glands are around the lip of the pitcher; in *Sarracenia* glands of this type may form an "ant-guiding" trail up the outer surface of the pitcher.

In the butterworts the glands that secrete the viscous globules of the "flypaper" leaf surface are specialized for that function only. In the sundews, however, the stalked glands secrete not only the adhesive but also the enzymes that digest the captured prey. The sundews also possess many minute stalkless glands, which are visible only with the microscope; these glands are scattered over the upper surface of the leaf and on the stalks of the larger glands. The function of the stalkless glands is not known, but it seems possible they are concerned with the movement of the larger stalked glands. The

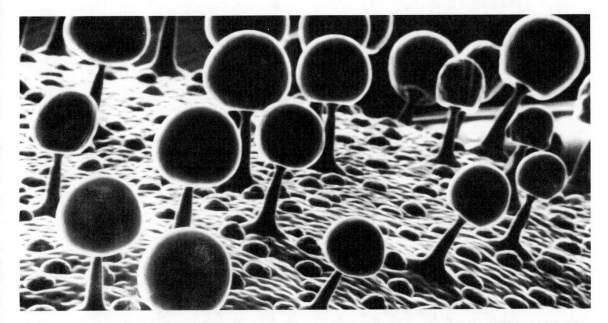

Figure 46 ARRAY OF GLANDS in the leaf surface of a butterwort, *Pinguicula grandiflora*, appears in this scanning electron micrograph. The prominent stalked glands trap insect prey. The many small semirecessed glands spread over the leaf surface are the digestive glands; after an insect has been captured they emit enzyme-rich fluid that forms a pool. The same glands later resorb the digestion product.

larger glands, which earlier observers saw as "tentacles," move in the direction of the prey when they are stimulated. The movement results from the loss of turgor in groups of cells along the side of the stalk closest to the stimulus. The stalkless glands may be responsible for the withdrawal of fluid that causes the loss of turgor.

Glands of a similar type may serve the same function in the butterworts when the leaf rolls up to enfold a captured insect, forming what Darwin called a temporary stomach. Evidence that certain glands perform this kind of function in the bladderworts has been presented recently by P. H. Sydenham and G. P. Findlay of Flinders University in Australia. The glands involved are on the outside of the bladders, and Sydenham and Findlay showed that the glands are concerned in the active transport of ions from inside the cavities of the traps to the bathing water outside. By setting up an osmotic gradient the ion movement generates an outward flow of water from the interior of the bladder. In the bladderworts such an outward flow is needed for the resetting of the trap. In the related genus *Genlisea* the prey moves along what is virtually a digestive tract, on the outside of which are glands similar to those found on the outside of the *Utricularia* bladder. Here they probably function to generate a flow of fluid along the tract.

The digestive glands of the carnivorous plants function under different conditions in the various genera according to the nature of the trapping mechanism. In *Nepenthes* the glands in the lower third of the pitcher become totally immersed in their own secretion fluids as the trap matures and before any prey is captured. In the larger species the accumulation of fluid can be as much as a liter. In the four other genera of pitcher plants smaller quantities of fluid collect, in some cases scarcely enough to immerse the glands, but here again it seems that the presence of prey is not necessary to stimulate secretion.

Figure 47 "TEMPORARY STOMACH" of the butterwort, as Charles Darwin called it, is seen unrolling in this photograph after the digestion of three fruit flies. No enzyme that can digest the chitinous external skeleton of the flies is present in the digestive fluid of butterwort glands, so that the bodies of the flies remain as empty shells after their inner tissues have been digested.

In contrast, the digestive glands on the leaf of the Venus's-flytrap remain dry until the prey is captured. If the trap is sprung with a pencil or a glass rod, the digestive glands remain dry and the leaf soon reopens. When an insect is trapped, however, the glands become active and a secretion pool builds up between the closed lobes of the leaf. Evidently the onset of secretion depends on chemical stimulation rather than mechanical. Although the glands in the bladders of the bladderworts are permanently immersed in water, it seems that they too do not secrete enzymes until they are stimulated by prey.

The sundews are different. Here the viscid secretion droplets accumulate on each gland head as the leaf matures; the load is held until the gland is touched by prey, and then still more secretion is released. The butterworts and the dewy pine *Drosophyllum* in some respects combine the features of the Venus's-flytrap and the sundews. In these genera there are two classes of glands on the leaf surface: stalked glands that carry secretion droplets at maturity and are mainly concerned with catching prey, and stalkless ones that remain dry until they are stimulated, when they pour out a less viscid secretion containing the digestive enzymes.

In Darwin's experiments on the sundews and butterworts he found that secretion could be stimulated by many sources of combined nitrogen but not, for example, by sugar or sodium carbonate. Insects excrete many nitrogenous products, but it has recently been suggested by Richard Robbins of the University of Oxford, who was studying the Venus's-flytrap, that the main stimulant may be uric acid, which is abundantly present in all insect excreta.

The digestive glands of the carnivorous genera vary considerably in their morphology. In the pitcher-plant genus *Nepenthes* the glands are some 60 micrometers in diameter and are partly sunk below the inner epidermis of the pitcher, where they are protected by an overlying flap of tissue. In the sundews the heads of the digestive glands are borne on multicellular stalks. The glands of the butterworts, both the stalked glands specialized for insect capture and the sessile ones concerned with digestion, are smaller and consist of many fewer cells than the glands of the other genera.

Notwithstanding such structural variations, a common architectural theme can be traced in all the digestive glands. Indeed, the theme is one that

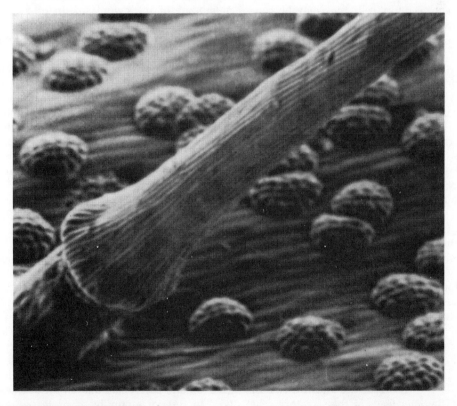

Figure 48 BASE OF A TRIGGER HAIR on the surface of a Venus's-flytrap leaf appears in this scanning electron micrograph. The low mounds are the glands that secrete the digestive fluid and then resorb the digestion products after the trap closes. The flow of secretion does not automatically follow the springing of the trap but depends on chemical stimuli provided by prey.

recurs in many other classes of plant-surface glands. In all cases it is the secretory cells that form an outer cap or layer one cell or a few cells thick, which lies directly over a specialized single cell or a pavement composed of several such cells side by side. This second layer is either in direct contact with the conducting vessels of vascular tissue or is separated from such tissue by two or three large "reservoir" cells.

The secretory outer cells of the gland are epidermal cells specialized for their function of enzyme synthesis, and they show many features reminiscent of those found in animal cells with similar functions. The network of cytoplasmic membranes known as the endoplasmic reticulum is well developed, and sometimes the elements are stratified, as they are in the cells of the animal pancreas. The endoplasmic reticulum is sometimes associated with colorless plastids. Plastids are a type of organelle not found in animal cells, the most familiar example being the green chloroplast of the photosynthetic apparatus. In the secretory cells they may be concerned in some manner with the synthesis of protein, but so far there is little evidence on this point.

The secretory cells of carnivorous plants are comparable to those of animals in still other ways. The vacuoles of the secretory cells, formed as inflated bays of the endoplasmic reticulum, are sites of enzyme storage; they are therefore comparable to the lysosomes of animal cells. Furthermore, in some instances the nuclei of the secretory cells of the gland head contain more DNA than most body cells. This is a feature of such animal glands as the salivary gland of the fruit fly. And finally one can see a parallel in the luxurious development of the Golgi apparatus in those glands concerned with mucilage secretion on the leaves of the sundews and butterworts. Many types of animal gland show a similar development of this cytoplasmic system, which is concerned with the packaging of various synthetic products and their passage out of the cell.

The cells of animal glands have an outer membrane but do not have a cell wall of the kind found in plants, and many of the adaptations of the gland cells of the carnivorous plants are unique in that they have to do with the structure and function of the cell wall. The outer walls of the secretory cells are coated with a water-resistant, cutinized layer, but this layer is perforated by distinct pores or less well-defined discontinuities through which the se-

cretions can reach the outer surface of the cell. The walls themselves are often modified for the storage and transfer of secretion products. Some are thickened irregularly to form extensive embayments or labyrinthine ramifications; the cell membrane follows these convolutions, so that the interface between the cell wall and the cytoplasm is greatly increased in area. For example, in the butterworts the interface may be enlarged by at least one order of magnitude. In the mucilage-secreting cells of the glands of the butterworts and the sundews the precursor products accumulate in the vesicles of the Golgi apparatus and discharge outward by fusing with the cell membrane, whence they pass across the cell wall and accumulate on its outer face. The vigor of this activity can be judged from the fact that the glands secrete several times their own volume of the mucilage during their active life.

The digestive glands secrete enzymes by other methods. In some instances the enzymes seem to diffuse directly through the plasmalemma, the outer membrane of the cytoplasm. In others, as in the sundews, the transfer involves a local disruption of the plasmalemma during the period of rapid secretion that follows the capture of the prey.

The cells of the layer underlying the secretory cells show some of the characteristics of those of the endodermal layer of the root, a sheath of cells that separates the root cortex from the inner conducting tissues. The side walls of the cells are heavily cutinized, and in them the plasmalemma is fused with the cell wall. Water cannot pass through the side walls, and so it is constrained to move through the cytoplasm.

Because the glands of the different genera of carnivorous plants function under widely different circumstances it is only to be expected that the processes of secretion and resorption should vary accordingly. My own observations of such "flypaper" trappers as butterworts and sundews suggest that these plants have secretory and resorptive mechanisms that are quite different from the ones likely to operate in pitcher plants. Among the butterworts some enzymes, notably amylases, are secreted by the stalked glands whose sticky exudate captures the insect prey, but it is the stalkless glands at the surface that furnish the main outflow of digestive fluid. Before stimulation the stalkless glands hold in reserve a supply of proteases, nucleases, phosphatases, esterases and other digestive enzymes, stored both in the spongy cell walls and in the vacuoles of the secretory cells. Stimulation induces an outpouring of fluid, and this flushes the stored enzymes out onto the surface of the leaf.

One can follow the activity of the enzymes that build up in the pool of secretion on the leaf surface. The pool extends and deepens to engulf the prey, and then, after the digestion is completed, the fluid is resorbed. Generally speaking, the size of the pool is related to the size of the prey. A small captive

Figure 49 DROSOPHILA in the process of being trapped by one of the several kinds of plants that augment their supply of nutrients by digesting animal life. The plant, *Drosera rotundifolia,* is one of the sundews. The stalks rising from the surface of the plant bear droplets of mucilage that hold the insect. The stimulus of contact makes adjacent stalks lose their turgor selectively and bend toward the insect, tying it down more securely. The same glands that secrete mucilage then exude digestive enzymes and later resorb the digestion products. (Photo by Thomas Eisner.)

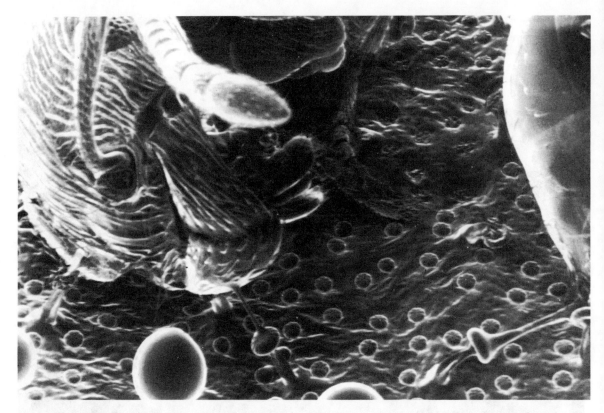

Figure 50 TRAPPED ANT is firmly attached to the leaf surface of a butterwort, held by strands of the mucilage secreted by the stalked glands of the leaf. Untouched stalked glands appear in the foreground of this scanning electron micrograph. The ant's head (*left*) has one of the adhesive strands attached to a mouthpart and another attached near the right antenna; a third strand is attached to the thorax (*upper right*) and two others secure the tip of the abdomen (*lower right*). The smaller, stalkless digestive glands have not yet started to secrete.

insect induces only a modest flow of digestive fluid, and in the butterworts after such a catch the pool may reach its maximum size in an hour or so. A large insect may stimulate so much secretion that surplus fluid will drip off the edges of the leaf. Under such circumstances the secretion may go on for several hours, and its volume may exceed the entire volume of liquid originally held in the leaf, showing that the flow is supplemented by the passage of water from elsewhere in the plant through the vascular system. Overstimulated leaves do not complete the digestive cycle. Resorption does not take place, and the leaf begins to rot, a victim, so to speak, of plant indigestion.

If the digestive cycle is normal, the period of resorption is only a little longer than the period of initial outflow. Knox and I found in our tracer experiments that the labeled end products of the di-

gestion of algal protein passed from the secretion pool back into the leaves of butterworts through the same glands that had supplied the digestive fluid. The spongy gland walls where the digestive enzymes had been stored became the channels for the inward passage. The products of digestion then passed through the endodermal cells into the vascular system of the leaf, where autoradiography detected them in the xylem vessels.

In other experiments we added a marker substance, colloidal lanthanum nitrate, to the secretion pools on butterwort leaves just as resorption was beginning. The marker is opaque under the electron microscope. As in the experiments with radioactive tracers, the substance could be tracked entering the gland cells through discontinuities in the cutin layer of the cell walls and moving on into the endodermal cells.

The protoplasts of the butterwort gland cells show striking changes during the process of digestion. In the secretory half of the cycle the vacuoles shrink and eventually become ill-defined. At the same time the plasmalemma, which is normally in continuous contact with the sinuous inner surface of the cell wall, loses this contact. As the glands enter the resorption phase of the cycle the cytoplasm condenses, the nucleus of the cell becomes clumped and the labyrinthine invaginations of the cell walls become blurred, probably through a partial dissolution of the wall structure.

In a detailed electron-microscope study of Venus's-flytrap glands during the digestive cycle D. Schwab, E. Simmons and James Scala of the Owens-Illinois Corporate Technology Technical Center in Toledo, Ohio, found corresponding changes in the fine structure of the cell walls. In the secretory half of the digestive cycle the invaginations of the cell wall became eroded and the plasmalemma took on a smoother profile. In the butterwort, where each gland functions only once, the

Figure 51 ANATOMY OF THE GLANDS of the dewy pine is shown in this drawing based on the original study by C. A. Fenner. Both the stalked glands that secrete the adhesive trapping mucilage and the stalkless glands that release enzymes after prey is trapped and then resorb products of digestion are connected directly with leaf's vascular system, or system of conductive vessels.

changes are irreversible. In the Venus's-flytrap, however, the walls of the secretory cells can be rebuilt and the cells can return to their former state before the leaf reopens to catch new prey.

The butterworts, and probably other genera of carnivorous plants with the same pattern of digestive-gland function, have thus evolved a definite digestive cycle. The secretion and resorption phases, respectively associated with massive movements of fluid first outward and then inward through each gland, are geared to far-reaching changes in the gland cells. There appears to be no such cycle in the pitcher plants. In this group the early period of secretion is followed by a prolonged interval when the pitcher holds a more or less constant amount of fluid. Trapped insects accumulate in the fluid, and the digestion products are withdrawn continuously from the pitcher into the main body of the plant. In a number of ingenious experiments Ulrich Lüttge of the Darmstadt School of Technology in West Germany has demonstrated that the same glands responsible for the secretion of pitcher-fluid enzymes also participate in the uptake of the digestion products. The process is essentially the same as in the butterworts except that resorption is achieved not through a wholesale uptake of fluid but through the selective inward passage of specific molecules and ions.

Lüttge found that the rates of uptake differ with the substance involved. For example, the amino acid alanine moved into the gland faster than phosphate ions, and phosphate ions were transported faster than sulfate ions. To explain these varying rates it is necessary to assume that participating in the uptake are "pumps" with different specificities. The pumps would be driven by the metabolic processes of the plant, and such an assumption is supported by the fact that transfer is partially paralyzed when metabolic inhibitors are present. Just where the pumps are located is not known, but it is significant that the large digestive glands of the *Nepenthes* pitcher possess the equivalent of an endodermal layer. Here, because of the thickening of the side walls of the cells, the fluids must move through the protoplasts. The metabolically driven transport systems may be incorporated in the membranes of the endodermal cells, if not in the membrane of the secretory cells themselves.

In the butterworts it is scarcely possible to explain the events of the digestive cycle in terms of selective pumping. The system involves mass flow in both

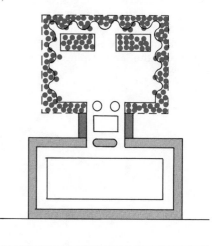

Figure 52 DIGESTIVE SEQUENCE is traced diagrammatically, based on the author's studies of butterworts. At the left is a digestive gland in its resting phase. Colored dots in the rectangular upper vacuoles and in the invaginations of some of the cell walls represent the stored digestive en-

directions. I have suggested that the initial outflow of digestive fluid following the capture of the prey is driven osmotically. If one supposes the stimulus of capture induces a rapid breakdown of the cell-wall polysaccharides, this could be the source of soluble sugars that could promote rapid fluid transfer into the gland cells by osmosis. In the initial period this flow would be through the intact membranes of the endodermal cells, water being abstracted first from the adjacent reservoir cell and then, through the contiguous vascular elements, from the rest of the plant. One can visualize the flow's reversing at the end of the secretion phase because control is lost by the now irreversibly altered endodermal cells. Resorption from the leaf surface would then be a matter of reverse flow through the gland and into the vascular system in response to diffusion gradients set up elsewhere in the plant.

The uptake mechanism Lüttge has proposed for the *Nepenthes* pitcher is distinctly similar to the mechanism assumed to be responsible for the nor-

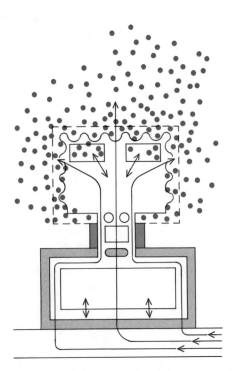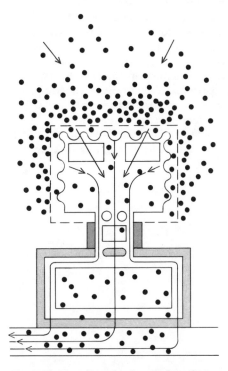

zymes. In the center the stimulus of prey capture induces an osmotically driven outward flow of fluid. This flow flushes out the stored digestive enzymes, which reach the surface of the leaf through discontinuities in the cuticle, the otherwise impermeable layer coating the cell wall. At

the right, after digestion is complete, the secretion pool on the surface of the leaf is resorbed, and the products of digestion (*black dots*) are transported through the cell wall and are distributed to the other parts of the plant through the plant's vascular system.

mal uptake of soil minerals by plant roots. It is as though in each pitcher the plant were creating its own enriched soil solution and abstracting from it the minerals it needs. The analogy seems even apter when one considers that after the pitcher has been open for some time its fluid becomes infected with a commensal flora, mostly bacteria, that quickly assumes most of the burden of digesting the captured prey.

At this stage the pitcher fluid has become distinctly akaline—and distinctly malodorous. The plant enzymes may now play little part in the digestive process; the digestive glands act mainly as organs of absorption, selectively taking up and concentrating useful products. To carry the analogy even further, it is possible that pitcher plants rooted in the ground may benefit at the beginning of each season's growth from a temporary local enrichment of the soil by nutrients released by the decay of the preceding season's dead pitchers and their partially digested contents. Here the useful products of pre-

dation would be taken up in the usual plant fashion: through the roots rather than the leaves.

It seems clear that the supplemental nutrients available to carnivorous plants offer them special advantages, particularly in environments where certain kinds of nutrients are scarce. It has commonly been supposed the principal benefit of the capture and digestion of animal prey by a plant is a supplemental supply of nitrogen. Current research indicates, however, that supplementary phosphorus is equally important and perhaps in some circumstances even more important. The presence in the digestive-gland secretions of nuclease and phosphatase enzymes may well be related to this requirement. In habitats where plant growth is limited by deficiencies of major nutrient elements such as phosphorus—or of one or more of the other elements required only in trace amounts—the advantages to be gained by acquiring contributions from animal prey would be substantial.

So much for the advantages of the carnivorous habit. Are there counterbalancing costs? Most plants live in competitive circumstances; is the carnivorous plant's energetic investment in the synthesis of digestive enzymes and other secretion products, not to mention the investment in the plant's elaborate structural adaptations, cost-effective? The intriguing conclusion of this line of thought is simply that any energy balance sheet is scarcely relevant. In all but a few instances the carnivorous plants are found in places where an abundance of sunlight, adequate carbon sources and unlimited access to water during the growing period place no limit on photosynthesis, the primary energetic resource of the plant. Thus the energetic cost of capturing an atom of nitrogen or of phosphorus, or of whatever else may be the principal growth-limiting element, is not significant. If the capture of such vital nutrients enables the plant to survive in places where no non-carnivorous competitor can intrude, then it is proved that, whatever the energetic cost may be, the investment is justified.

THE PERILS OF PARASITISM

. . .

Introduction

. . .

Symbiosis, a mutually beneficial interaction between two species, has been a recurrent theme throughout this volume: eukaryotic cells began as symbiotic corporations of bacteria, sulfide-metabolizing bacteria share resources with the vent worms that house them, koalas depend on the cellulose-digesting microorganisms in their guts, and so on. Many symbioses are obligate and so neither partner can survive without the other; others are facultative—good for both but not essential to either. Other interactions are less even: commensual species benefit from their association with another, but provide nothing in return, and parasites actually harm their hosts.

Most parasites are obligate: they must find a victim or perish. And having found one victim, they must be able to move to another (or get their offspring to a new host) before their present living habitat dies. Some parasites require hosts of a single species; others must infect three or four different hosts in a specific order. The Chinese liver fluke, for example, lays its eggs in the human liver. The eggs pass via the bile to the intestines and then out with the feces; at that point the eggs must be eaten by a particular species of freshwater snail, in which they hatch and grow. When they leave the snail the parasites must find a fish, burrow in and encyst in the host's muscles; the fluke completes its cycle when humans eat infected fish that has been insufficiently cooked.

In Chapter 9, "Communication between Ants and Their Guests," we see a relatively benign class of parasitism. By breaking the chemical and tactile codes of particular species of ants, other sorts of insects are able to gain access to the protection of the nest, where they are fed and tended by their hosts (often in preference to the ants' own young) and even eat the hosts' own larvae. The "guests" of army ants mimic the appearance of their hosts, thereby gaining protection from insectivores that have learned to avoid these ferocious marauders. Given that most species of army ants are actually blind, the mimicry is unlikely to be for the purpose of deceiving the hosts.

Even ants parasitize ants. Chapter 10, "Slavery in Ants," describes two sorts of parasitism: true slavery and nest parasitism. The workers of slave-making ants carry off the young of another species, which, when they metamorphose into adults, take up the duties of brood rearing, nest maintenance and foraging. The workers of obligate slave makers can do nothing but raid other colonies for a continuing supply of impressed labor.

Nest parasites, on the other hand, live in the host colony. The queens of some species invade the nest of the host species and kill the workers and the queen. When the host larvae and pupae mature, they begin caring for her and rearing her young. The queens of other species insinuate themselves into the hosts' nests and hide themselves until they pick up the general colony odor. That chemical invisibility allows them to search with impunity until they find the host queen and kill her. This strategy allows the parasite to start off with a complete complement of workers and young.

In either case, however, the parasitized colony is doomed: once the existing supply of host eggs has matured and lived out the normal worker lifespan, the parasites are unable to take over the domestic chores, and the nest dies out. But before that fate overtakes it, the parasite queen produces thousands of eggs that the hosts rear dutifully into reproductives; the new queens mate and then leave in search of new colonies to parasitize.

One species has its cake and eats it too. *Teleutomyrmex* (literally the "ultimate ant") queens insinuate themselves into a host nest and find the queen, but instead of killing her they climb onto the host's back or abdomen and attach themselves. The parasitic queens are fed by the host workers and lay their eggs along with the host queen. The mixed broods are reared together, but all the parasite's eggs turn into reproductives that, when adults, mate within the nest. The new queens fight for space on the host's body; unsuccessful queens leave in search of other colonies. These parasitized nests live on indefinitely with a steady supply of host workers to tend the parasites.

Figure 53 QUEENS OF *TELEUTOMYRMEX*, the "ultimate ant," ride on the host queen and are fed by host workers (*lower right*) as they dispense their eggs along with those of the host queen. Their offspring, all reproductives, mate in the colony. The new queens must either fight for a place on the host queen or leave in search of a new colony. (Painting by H. Kutter as photographed by O. Bauer, courtesy of the Department of Library Services, American Museum of Natural History.)

The most specialized parasites of all are internal parasites. At first glance they would seem to face the worst possible odds: host animals—mammals in particular—have an elaborate immune system dedicated specifically to killing foreign organisms. Because of the way the immune system works, however, novel invading parasites have a few days of grace. During early development the body creates millions of different B cells, each with its own unique kind of antibody on its cell surface. When a foreign organism enters the circulatory system, at least one of the antibody types is likely to bind to a protein in the parasite's cell coat; those proteins are called antigens. The binding triggers the self-cloning of the B cell in question, creating thousands of identical antibody-producing cells within a week and a full-fledged and highly specific immune response ensues [see "The Development of the Immune System," by M. D. Cooper and A. R. Lawton; SCIENTIFIC AMERICAN, November, 1974; Offprint 1306]. Subsequent infections by the same parasite are dealt with much faster because the initial priming has already taken place—so fast, in fact, that we usually have no symptoms of infection and are said to be "immune."

Chapter 11, "How the Trypanosome Changes Its Coat," describes how the microorganism that causes sleeping sickness beats the immune system by switching its antigens before the immune response is fully under way, so that the host's battle forces are always just a bit too late. The parasite's trick is to move one of its thousands of alternative coat-protein genes into a special part of the trypanosome chromosome whenever the going is getting tough; such movable genes are known as transposons [see "Transposable Genetic Elements," by Stanley N. Cohen and James A. Shapiro; SCIENTIFIC AMERICAN, February, 1980]. The irony is that the trypanosome is beating the immune system by using the same strategy that the B cells use in generating the antibody diversity that allows them to respond to invaders in the first place: during development the unique DNA sequence coding for a cell's antibody is constructed in a cut-and-paste fashion out of a set of alternative transposable genetic segments [see "The Genetics of Antibody Diversity," by Philip Leder; SCIENTIFIC AMERICAN, May, 1982].

We began this volume with the origin of life, with the development of order and complexity out of the perilous chemical chaos of the early earth. Although our everyday picture of the world makes it appear that living things have, on the whole, become steadily larger and more sophisticated with the passing millennia, natural selection has in fact simply favored whatever works, be it big or small, crude or complex. The utterly pragmatic nature of evolution is beautifully illustrated in Chapter 12, "Viroids," a discussion of "organisms" so pointless and degenerate that they make viruses seem elegant. Viroids consist of nothing more than naked bits of RNA that insinuate themselves into their hosts and induce the production of more of the same sequences. The RNA does not even encode an enzyme or other protein; it simply contains some recognition sequences that facilitate its transport and replication, and would have gone unnoticed were it not that a few kinds cause visible damage to economically important plants. Viroids, surely, represent life so near the edge that we can debate whether, even though it reproduces itself, it is life at all.

There may be an even less believable class of parasite, consisting only of a protein fragment, that somehow induces the production of more of the same [see "Prions," by Stanley B. Prusiner; SCIENTIFIC AMERICAN, October, 1984; Offprint 1554]. An agent of that sort is responsible for scrapie, a degenerative disease of the central nervous system of sheep and goats that had been attributed to viroids. It is likely that each of us is infected with countless genetic "parasites" that are for the most part benign: there are thousands of so-called "repetitive sequences" of DNA in the chromosomes that serve no purpose for us, their hosts, but ensure their own continued perpetuation. Given that the vast majority of the DNA of higher animals is never transcribed into RNA, it may be that the genetic parasites accumulated over three billion years of evolution now outnumber the genes of their unwilling and unsuspecting hosts.

Communication between Ants and Their Guests

Ants feed and shelter many other species of arthropods. The key to this hospitality lies in the guests' ability to communicate in the same chemical and mechanical language used by their hosts.

. . .

Bert Hölldobler

March, 1971

The world of the arthropods is marked by a curious phenomenon, discovered nearly a century ago, that has never ceased to intrigue investigators. Many species of insects and other arthropods live with ants and have developed a thriving parasitic relationship with them. A number of these myrmecophiles make their home in the ants' nests and enjoy all the benefits. Although the interlopers in some cases eat the host ants' young, the ants treat the guests with astonishing cordiality: they not only admit the invading species to the nest but feed, groom and rear the guest larvae as if they were the ants' own brood.

How do the myrmecophiles manage to gain this acceptance? Ants, as highly social animals, possess a complex system of internal communication that enables the colony to carry out its collaborative activities in nest-building, food-gathering, care of the young and defense against enemies. The fact that the ants do not treat their alien guests as strangers suggests that the guests must somehow have bro-ken the ants' code, that is, attained the ability to "speak" the ants' language, which involves a diversity of visual, mechanical and chemical cues.

Studies of the social behavior and communication of ants over the past 10 years have produced information that now provides a basis for well-grounded investigation of the relations between myrmecophiles and their hosts. Among the thousands of myrmecophilous animals (which include not only such arachnids as mites but also collembolans, flies, wasps and many other insect groups) the staphylinid beetles, commonly known as the rove beetles, demonstrate the parasitic relationships in a particularly clearcut fashion. Focusing mainly on this family of myrmecophiles, I have been looking into the details of their communication and relations with certain species of ants.

The relations vary considerably with the beetle species. Some species live along the ants' food-gathering trails, others at the garbage dumps outside the nest, others in outer chambers within the nest and still others all the way inside the brood chambers. Let us consider first the beetles that attain the brood chamber.

A well-known example of such a beetle is *Atemeles pubicollis,* a European species. During its larval stage it lives in the nest of the mound-making wood ant *Formica polyctena.* I found that the ant's adoption of the beetle larva depends in the first instance on chemical communication. The larva secretes from glandular cells in its integument a substance that apparently acts as an attractant for the ant. This substance may be an imitation of a pheromone that ant larvae themselves emit to release brood-keeping behavior in the adults [see "Pheromones," by Edward O. Wilson; SCIENTIFIC AMERICAN, May, 1963]. The brood-tending ants respond to the chemical signal from the *Atemeles* beetle larvae with intense grooming of the larvae. I was able to verify

the existence of chemical communication by two kinds of experimental evidence. Experiments with radioactive tracers demonstrated that substances were transferred from the beetle larvae to the ants. When larvae coated with shellac to prevent liberation of their secretion were placed at the entrance to the nest, the ants either ignored the larvae or carried them off to the garbage dump. If at least one segment of a larva's body was left uncovered by shellac, however, the ants would take the larva into the nest and adopt it. They even carried in dummies of filter paper soaked with secretions extracted from the beetle larvae.

A different form of communication comes into play to elicit the ants' feeding of larvae. The beetle larvae imitate a certain begging behavior of ant larvae involving mechanical stimulation of the

Figure 54 YOUNG GUEST, a beetle larva of the genus *Atemeles,* is given a droplet of liquid food by an ant attendant in the brood chamber of a nest of *Formica* ants. Not only beetles but also wasps, flies and many other arthropods are fed and sheltered by various ant hosts.

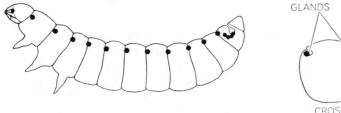

GLANDS

CROSS
SECTION

1

2

3

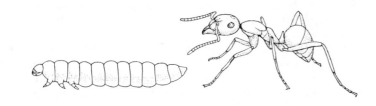

4

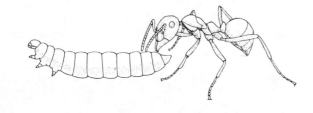

5

Figure 55 CHEMICAL ATTRACTANT secreted by a row
of glands on each side of the beetle larva (*top*) causes the
ant host to groom the larva intensively (*1 – 4*). Experiments
with tracer substances indicate that the grooming process
transfers secretions from larva to ant. Contact between ant
and larva also makes the larva rear up. If the larva can
make mouth-to-mouth contact with the ant (*5*), the stimulus causes the ant to regurgitate a droplet of liquid food.
With the exception of Figure 62, the representations of ant
hosts and their guests are based on a series of original
illustrations by Turid Hölldobler.

brood-keeping adults. When a larva is touched by the adult ant's mouthparts or antennae, it promptly rears up and tries to make contact with the ant's head. If the larva succeeds in tapping the ant's lip with its own mouthparts, the ant regurgitates a droplet of food. The beetle larvae perform the begging behavior more intensely than the ant larvae do, and apparently for this reason they receive more food. In order to trace and measure the distribution of food to the larvae in a brood chamber I gave the ants food labeled with radioactive sodium phosphate. The experiment showed that in a mixed population of beetle and ant larvae the beetles obtained a disproportionately large share of the food. The presence of beetle larvae reduced the normal flow of food to the ant larvae; on the other hand, the presence of ant larvae did not affect the food flow to the beetle larvae.

The question arises: How does the ant colony manage to survive the beetle larvae's competition for the food and their intense predation on the ant larvae in the brood chamber? This turns out to have a simple answer. The beetle larvae are cannibalistic and unable to distinguish their fellow larvae from ant larvae by odor. They therefore cut down their own population, whereas the ant larvae do not. Typically one finds that in a brood chamber the ant larvae are found in clusters but beetle larvae (particularly those belonging to the genus *Lomechusa*) show up as loners, having devoured their neighbors.

The *Atemeles* beetles have two homes with ants —one for the summer, the other for winter. After the larvae have pupated and hatched in a *Formica* nest, the adult beetles migrate in the fall to nests of dark brown, insect-eating ants of the genus *Myrmica*. The reason for the beetles' migration is that brood-keeping and the food supply are maintained in *Myrmica* colonies throughout the winter, whereas *Formica* ants suspend their raising of young. In the

Figure 56 MATURING LARVA obtains added nourishment in the *Formica* brood chamber by eating the small ant larvae sheltered there. The beetle larvae also prey on one another. The nest in this photograph and the one in Figure 54 are man-made laboratory structures.

Myrmica nests the beetles, still sexually immature, can be fed and ripen to maturity by the spring, at which time they return to *Formica* nests for mating and the laying of eggs. Thus the life cycles and behavior of the *Atemeles* beetles and the *Formica* and *Myrmica* ants are so synchronized that the beetles can take maximum advantage of the social life of each of the two ant species that serve as hosts. In this respect *Atemeles* shows a remarkably advanced form of evolutionary adaptation. The *Lomechusa* beetles, also co-dwellers with *Formica* ants, do not change their environment for the winter; after hatching they simply move on to another *Formica* colony of the same species and share the food shortage. It appears that *Atemeles* evolved myrmecophilous relations with *Formica* to begin with and then "discovered" and adapted to a winter home with *Myrmica*, developing proficiency in a second language for that purpose.

Before leaving the *Formica* nest at the beginning of the fall the *Atemeles* beetle obtains a supply of food for its migration by begging from its *Formica* hosts. For this it employs a technique of tactile stimulation. The beetle first drums rapidly on an ant with its antennae to attract attention. It then induces the ant to regurgitate food by touching the ant's mouthparts with its maxillae and forelegs. High-speed motion pictures show that ants themselves obtain food from one another by similar mechanical signals.

How does the migrating beetle find its way to a *Myrmica* nest? *Formica* nests are normally found in woodland, whereas *Myrmica* nests are found in the grassland beyond the woods. It can be shown experimentally that when *Atemeles* beetles leave the *Formica* nest, they generally move in the direction of increasing light. This may explain how the beetles manage to reach the relatively open grasslands where the *Myrmica* ants live. After reaching the open grassland the beetle has to use other cues to find a *Myrmica* nest. By means of laboratory experiments I have ascertained that it is guided to the nest by the odor of the host species of ant. The odor must be windborne; the beetle is not drawn to it in still air. Curiously, the beetle possesses only a temporary sensitivity to this odor; it is limited to the two weeks after leaving the *Formica* nest. (In the spring the beetle locates a *Formica* nest in the same way, by its odor.)

On finding a *Myrmica* nest, the beetle obtains recognition and adoption with a ritual involving

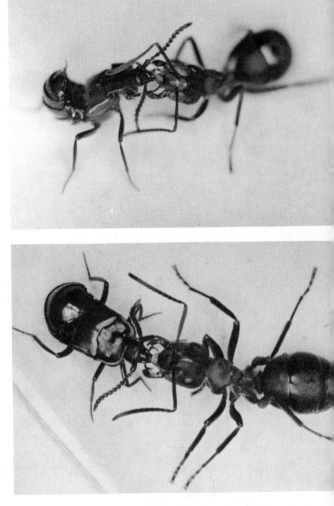

Figure 57 IDENTICAL MEANS are used by ants and beetles to make a food-laden forager regurgitate. A tapping of the forager's mouthparts with the food-seeking beetle's forelegs is the key stimulus (*top*). Regurgitation is immediate (*bottom*).

chemical communication. The beetle first taps the ant lightly with its antennae and raises the tip of its abdomen toward the host. The ant responds by licking secretions from glands on the abdomen's tip that I call "appeasement glands," because their secretion apparently suppresses aggressive behavior in the ant. The ant is next attracted to a series of glands along the sides of the beetle's abdomen; I call these the "adoption glands," as the ant will not welcome or adopt the beetle unless it senses their

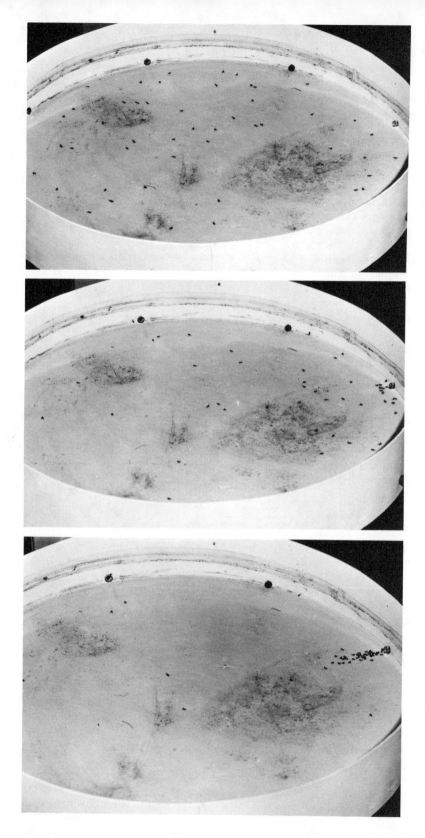

Figure 58 ROLE OF SCENT in directing the migration of *Atemeles* beetles from their initial residence in *Formica* nests to winter quarters in *Myrmica* nests is demonstrated in these laboratory photographs. At the start of the experiment (*top*) the beetles are distributed at random in a circular enclosure. Air scented with the odor of *Myrmica* ants is then blown into the enclosure through one of the eight holes around its rim (*center*). Ten minutes later (*bottom*) most of the beetles have collected in a cluster near the scent-emitting hole.

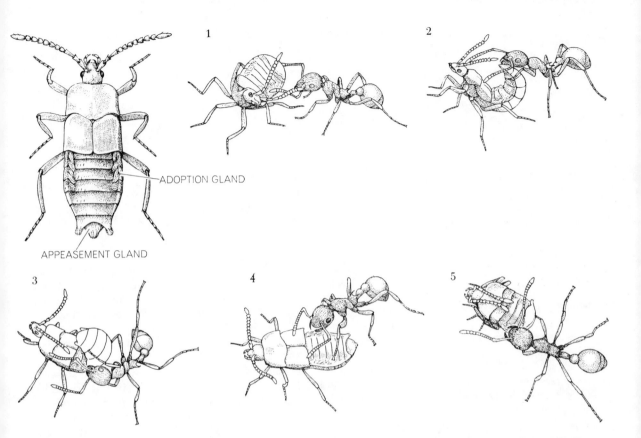

ADOPTION GLAND

APPEASEMENT GLAND

1

2

3

4

5

Figure 59 ADOPTION RITUAL, which gains entry for *Atemeles* beetles to nests of the ant genus *Myrmica*, depends on secretions from glands at the tip and along the sides of the beetle's abdomen (*left*). On encountering a potential *Myrmica* host in the nest antechamber, the beetle attracts its attention (*1*) by tapping the ant lightly with its antennae and raising the tip of its abdomen. The response of the ant (*2*) is to taste the secretion from the glands there, known as "appeasement glands" because they apparently suppress the ant's aggressive behavior toward intruders. The next attractant is the secretion from the "adoption glands" (*3, 4*), so named because an ant will not welcome an intruder before it senses this secretion. The tightly curled beetle is then carried to a brood chamber (*5*).

secretion. Presumably the odor of this secretion mimics the odor of the ant species. Finally the beetle lowers its abdomen so that the ant can approach, and the ant then grasps some bristles around the beetle's lateral glands and carries the guest into the brood chamber.

The *Atemeles* beetles are not the only myrmecophiles capable of making themselves at home with more than one kind of ant. More than 50 years ago the Harvard University entomologist William Morton Wheeler discovered that staphylinid beetles of the *Xenodusa* genus change their domicile with the seasons. The larvae live in *Formica* nests through the summer and the adults overwinter in nests of the carpenter ants of the genus *Camponotus*. It is interesting to note that the carpenter ants also maintain larvae throughout the winter. It may well

be that the evolutionary history of the *Xenodusa* beetles parallels that of *Atemeles* in selecting and adapting to a winter home.

Unlike the *Atemeles*, *Lomechusa* and *Xenodusa* beetles, various other genera of staphylinid beetles do not possess the command of ant language that is required to gain entry into the brood chamber, the optimum niche for obtaining food. Staphylinids of the European genus *Dinarda*, for example, are limited to the peripheral chambers of the nests of their host (*Formica sanguinea*). *Dinarda* offers secretions from glands similar to *Atemeles*' appeasement glands, but these secretions only induce the ant to tolerate the beetle, not to adopt it and take it into the brood chamber. *Dinarda* is therefore reduced to living on such food as it can find or scrounge in the peripheral chambers. There it feeds on dead ants

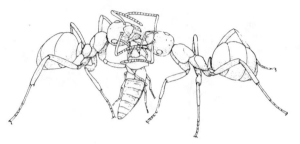

Figure 60 FOOD THIEVES, beetles of the genus *Dinarda*, are tolerated in the outer chambers but not the brood chambers of the *Formica* nest. By touching a forager ant's mouthparts (*left*) a *Dinarda* beetle may induce regurgitation of a food droplet. The beetle may also intercept a droplet (*right*) as it is being passed between two ants. If attacked, the beetle raises its abdomen, drawing attention to the glands that, like the *Atemeles* "appeasement glands," secrete a substance that the ants savor. While the ant tastes, the beetle flees.

that have not yet been taken to the garbage dump and on food that it may steal or wheedle from the worker ants. Occasionally *Dinarda* snatches a food droplet from the mouth of a forager that is about to pass the food to another worker. Or the beetle may surreptitiously approach a food-laden forager and, by touching the forager's lip, induce the regurgitation of a small food droplet. The ant immediately recognizes the beetle as an alien, however, and starts to attack it. The beetle staves off the attack by raising its abdomen and offering the ant its appeasement secretion; while the ant is savoring the substance it has licked up, the beetle makes its escape.

Other groups of myrmecophilous staphylinid beetles (for example the genus *Myrmedonia*) have a bare minimum of communication with their ant hosts, sufficient only to allow the beetle to feed at the ants' garbage dumps. At the dump the foraging beetle can avert attack by an ant from the nest by offering it the appeasement secretion and thus winning time to escape. If the beetle is placed anywhere inside the nest, however, the appeasement does not avail; the ants promptly kill the beetle as an intruder.

There are myrmecophiles that possess only an elementary, one-way form of chemical communication with the ants they depend on for food. This is simply the ability to recognize the odor of the trail laid down between the nest and a food source by foragers of the particular ant species. For example, a small nitidulid beetle in Europe (*Amphotis marginata*) can thus identify the trails of a shiny black wood ant (*Lasius fuliginosus*), and in many localities these beetles abound along the trails. They act as

begging highwaymen, intercepting food-laden ants on the trail and inducing them to regurgitate food droplets by tapping the ant's labium. The ant soon realizes it has been tricked and attacks the beetle. Although the beetle has no appeasement mechanism, it avoids injury by retracting its appendages and flattening itself on the ground.

Many myrmecophilous beetles closely resemble their ant hosts in appearance. This is particularly true of guests of the army ants, and some investigators concluded that the factor inducing the ants to accept beetles as their nestmates was the beetles' morphological resemblance to themselves. It was even thought to be the case with *Atemeles* and *Lomechusa*, although they do not particularly resemble their hosts. I have been able to show, by artificially altering the shape and color of these beetles, that morphological features do not contribute to the success of their relationship with their hosts. Instead it appears that communicative behavior remains the essential requirement for acceptance. I believe this is probably generally the case, even for the guests of the army ants. Very likely the guests' mimicry of their hosts' appearance has evolved as a protection against predation by birds. Predatory birds that follow army ants as they travel over open ground do not attack the ants themselves; they feed only on other insects that are stirred up by the ants' march. Presumably insects that resemble ants are ignored by the birds. With carefully designed experiments we should be able to resolve this question.

There remains for further exploration the fascinating question of how the extraordinarily effective system of communication between myrmecophiles and their hosts evolved. We can think of

1

2

3

4

Figure 61 HIGHWAYMAN BEE-
TLES of the genus *Amphotis* locate foraging trails of the wood
ant *Lasius* by the scent and am-
bush food-laden workers. By
stimulating the ant's mouthparts
(*1*) the beetle causes it to regurgitate its cropful of food (*2*). The
robbed ant then reacts aggres-
sively (*3*), but passive defense (*4*)
enables the armored beetle to
weather the attack.

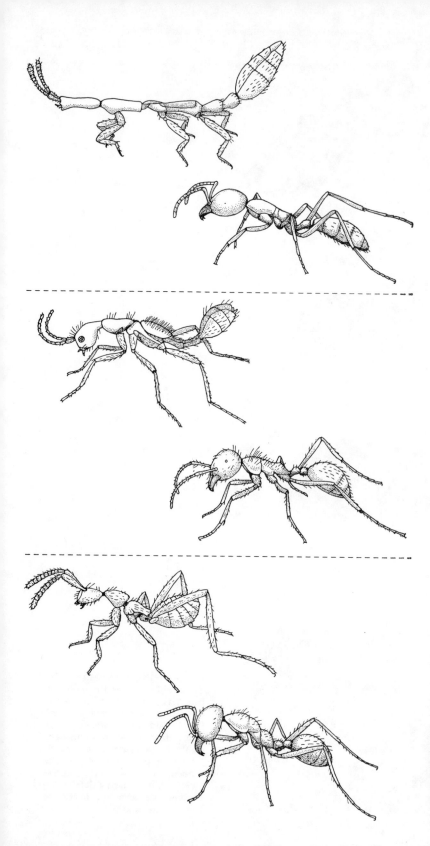

Figure 62 BEETLE MIMICS that resemble their army-ant hosts include a seemingly headless species, *Mimanomma spectrum (top left)*. Its host is the African driver ant *Dorylus nigricoms (top right)*. Other army-ant mimics are *Crematoxenus aenigma (center left)* and *Mimeciton antennatum (bottom left)*, shown with their respective New World hosts, *Neivamyrmex melanocephalum* and *Labidus praedator*. The drawings are based on original sketches by the late Charles H. Seevers of Roosevelt College. There is no evidence that mimicry affects guest-host relations. It may, however, protect the beetles from attack by other predators.

this evolution as a two-part process. First, viewing the myrmecophile as a signal-receiver, we can suppose the potential guest of a specific ant underwent gradual evolutionary modification of its receptor system that endowed it with the ability to recognize the ant's odor, the distinctions between the ant larva and the adult and other signals opening the way to a parasitic relationship. Second, regarding the myrmecophile as a transmitter of signals, we can see that through natural selection it must have evolved the set of secretions and behavioral acts that induces the ant to accept the guest in the nest and nurture it. Thus the development of the guest-host relationship would involve adaptive changes only in the guest, accommodating itself to the nature of the specific host. In all probability this is the way *Atemeles* and *Lomechusa* won their welcome and support in the homes of the respective species of ant hosts.

By careful analysis of various related species of beetles that enjoy differing degrees of intimacy with their hosts we can expect to learn more about the details of the evolution of the social associations between myrmecophiles and ants and their communication systems.

Slavery in Ants

Certain species of ants raid the nests of other species for ants to work in their own nest. Some raiding species have become so specialized that they are no longer capable of feeding themselves.

• • •

Edward O. Wilson

June, 1975

The institution of slavery is not unique to human societies. No fewer than 35 species of ants, constituting six independently evolved groups, depend at least to some extent on slave labor for their existence. The techniques by which they raid other ant colonies to strengthen their labor force rank among the most sophisticated behavior patterns found anywhere in the insect world. Most of the slave-making ant species are so specialized as raiders that they starve to death if they are deprived of their slaves. Together they display an evolutionary descent that begins with casual raiding by otherwise free-living colonies, passes through the development of full-blown warrior societies and ends with a degeneration so advanced that the workers can no longer even conduct raids.

Slavery in ants differs from slavery in human societies in one key respect: the ant slaves are always members of other completely free-living species that themselves do not take slaves. In this regard the ant slaves perhaps more closely resemble domestic animals—except that the slaves are not allowed to reproduce and they are equal or superior to their captors in social organization.

The famous Amazon ants of the genus *Polyergus* are excellent examples of advanced slave makers. The workers are strongly specialized for fighting. Their mandibles, which are shaped like miniature sabers, are ideally suited for puncturing the bodies of other ants but are poorly suited for any of the routine tasks that occupy ordinary ant workers. Indeed, when *Polyergus* ants are in their home nest their only activities are begging food from their slaves and cleaning themselves ("burnishing their ruddy armor," as the entomologist William Morton Wheeler once put it).

Figure 63 RAID BY SLAVE-MAKING AMAZON ANTS of the species *Polyergus rufescens* (*light color*) against a colony of the slave species *Formica fusca* (*dark color*) is depicted. The *fusca* ants make their nest in dry soil under a stone. The raiding Amazon ants kill resisting *fusca* workers by piercing them with their saberlike mandibles. Most of the Amazon ants are transporting cocoons containing the pupae of *fusca* workers back to their own nest. When the workers emerge from the cocoons, they serve as slaves. Two dead *fusca* workers that resisted lie on the ground. Two other workers have retreated to upper surface of the rock over the nest's entrance.

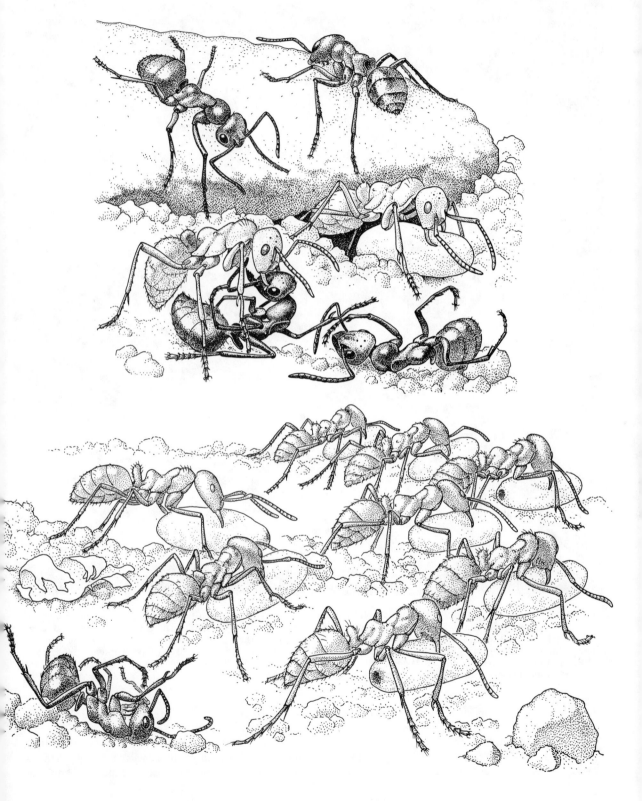

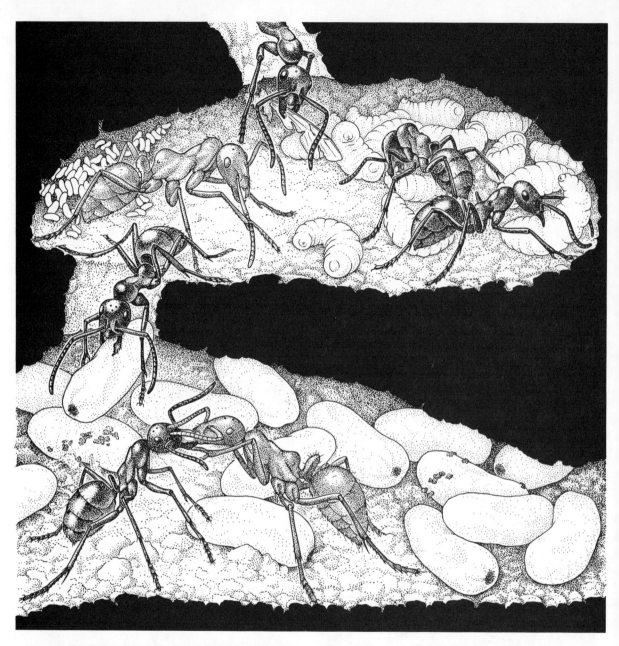

Figure 64 INTERIOR VIEW OF THE HOME NEST of a colony of Amazon ants shows *Formica fusca* slaves (*dark color*) performing all the housekeeping labor. At top center one of the slaves brings a fly wing into the nest for food. Other slave workers care for the small eggs, grublike larvae and cocoon-enclosed pupae of their captors. During the raiding season some of the pupae are likely to be those of *fusca* workers. The slave makers (*light color*) can do nothing more than groom themselves (*upper left*). In order to eat, the Amazon ants must beg slave workers to regurgitate liquid droplets for them (*lower left*). These ant species are found in Europe.

When *Polyergus* ants launch a raid, however, they are completely transformed. They swarm out of the nest in a solid phalanx and march swiftly and directly to a nest of the slave species. They destroy the resisting defenders by puncturing their bodies and then seize and carry off the cocoons containing the pupae of worker ants.

When the captured pupae hatch, the workers that emerge accept their captors as sisters; they make no distinction between their genetic siblings and the

Polyergus ants. The workers launch into the round of tasks for which they have been genetically programmed, with the slave makers being the incidental beneficiaries. Since the slaves are members of the worker caste, they cannot reproduce. In order to maintain an adequate labor force, the slave-making ants must periodically conduct additional raids.

It is a remarkable fact that ants of slave-making species are found only in cold climates. Although the vast majority of ants live in the Tropics and the warm Temperate zones, not a single species of those regions has been implicated in any activity remotely approaching slavery. Among the ants of the colder regions this form of parasitism is surprisingly common. The colonies of many slave-making species abound in the forests of the northern U.S., and ant-slave raids can be observed in such unlikely places as the campus of Harvard University.

The slave raiders obey what is often called Emery's rule. In 1909 Carlo Emery, an Italian myrmecologist, noted that each species of parasitic ant is genetically relatively close to the species it victimizes. This relation can be profitably explored for the clues it provides to the origin of slave making in the evolution of ants. Charles Darwin, who was fascinated by ant slavery, suggested that the first step

was simple predation: the ancestral species began by raiding other kinds of ants for food, carrying away their immature forms in order to be able to devour them in the home nest. If a few pupae could escape that fate long enough to emerge as workers, they might be accepted as nestmates and thus join the labor force. In cases where the captives subsequently proved to be more valuable as workers than as food, the raiding species would tend to evolve into a slave maker.

Although Darwin's hypothesis is attractive, I recently obtained evidence that territorial defense rather than food is the evolutionary prime mover. I brought together in the Harvard Museum of Comparative Zoology different species of *Leptothorax* ants that normally do not depend on slave labor. When colonies were placed closer together than they are found in nature, the larger colonies attacked the smaller ones and drove away or killed the queens and workers. The attackers carried captured pupae back to their own nests. The pupae were then allowed by their captors to develop into workers. In the cases where the newly emerged workers belonged to the same species, they were allowed to remain as active members of the colony. When they belonged to a different *Leptothorax* spe-

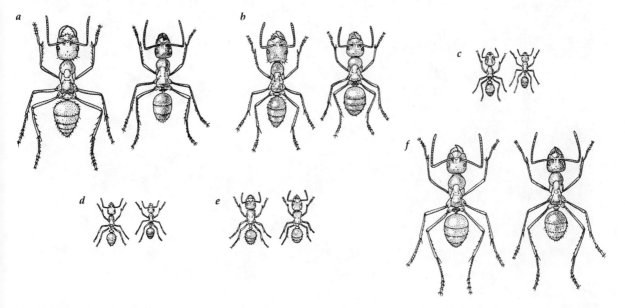

Figure 65 RESEMBLANCE of slave maker and slave was noted by an Italian myrmecologist, Carlo Emery, in 1909. In each pair of ants shown here the slave maker is on the left and the slave on the right. The species depicted are (a) *Polyergus rufescens* and *Formica fusca*, (b) *Rossomyrmex proformicarum* and *Proformica nasutum*, (c) *Harpagoxenus americanus* and *Leptothorax curvispinosus*, (d) *L. duloticus* and *L. curvispinosus*, (e) *Strongylognathus alpinus* and *Tetramorium caespitum* and (f) *F. subintegra* and *F. subsericea*.

cies, however, they were executed in a matter of hours. One can easily imagine the origin of slave making by the simple extension of this territorial behavior to include tolerance of the workers of related species. The more closely related the raiders and their captives are, the more likely they are to be compatible. The result would be in agreement with Emery's rule.

One species that appears to have just crossed the threshold to slave making is *Leptothorax duloticus*, a rare ant that so far has been found only in certain localities in Ohio, Michigan and Ontario. The anatomy of the worker caste is only slightly modified for slave-making behavior, suggesting that in evolutionary terms the species may have taken up its parasitic way of life rather recently.

In experiments with laboratory colonies I was able to measure the degree of behavioral degeneration that has taken place in *L. duloticus*. Like the Amazon ants, the *duloticus* workers are highly efficient at raiding and fighting. When colonies of other *Lep-*

tothorax species were placed near a *duloticus* nest, the workers launched intense attacks until all the pupae of the other species had been captured.

In the home nest the *duloticus* workers were inactive, leaving almost all the ordinary work to their captives. When the slaves were temporarily taken away from them, the workers displayed a dramatic expansion in activity, rapidly taking over most of the tasks formerly carried out by the slaves. The *duloticus* workers thus retain a latent capacity for working, a capacity that is totally lacking in more advanced species of slave-making ants.

The *duloticus* workers that had lost their slaves did not, however, perform their tasks well. Their larvae were fed at infrequent intervals and were not groomed properly, nest materials were carried about aimlessly and were never placed in the correct positions, and an inordinate amount of time was spent collecting and sharing diluted honey. More important, the slaveless ants lacked one behavior pattern that is essential for the survival of the colony: foraging for dead insects and other solid food. They even

Figure 66 COLONY OF ANTS housed in a glass tube consists of the rare species *Leptothorax duloticus* and a slave species, *L. curvispinosus*. The *duloticus* ant, found in Ohio, Michigan and Ontario, has only recently become a slave maker. One of the *duloticus* workers can be seen in the center of the photograph; below it are three slave workers. The white objects are immature forms of both species. When the slave workers are removed, the *duloticus* workers attempt to carry out necessary housekeeping tasks but do so poorly.

ignored food placed in their path. When the colony began to display signs of starvation and deterioration, I returned to them some slaves of the species *Leptothorax curvispinosus*. The bustling slave workers soon put the nest back in good order, and the slave makers just as quickly lapsed into their usual indolent ways.

Not all slave-making ants depend on brute force to overpower their victims. Quite by accident Fred E. Regnier of Purdue University and I discovered that some species have a subtler strategy. While surveying chemical substances used by ants to communicate alarm and to defend their nest, we encountered two slave-making species whose substances differ drastically from those of all other ants examined so far. These ants, *Formica subintegra* and *Formica pergandei,* produce remarkably large quantities of decyl, dodecyl and tetradecyl acetates. Further investigation of *F. subintegra* revealed that the substances are sprayed at resisting ants during slave-making raids. The acetates attract more invading slave makers, thereby serving to assemble these ants in places where fighting breaks out. Simultaneously the sprayed acetates throw the resisting ants into a panic. Indeed, the acetates are exceptionally powerful and persistent alarm substances. They imitate the compound undecane and other scents found in slave species of *Formica*, which release these substances in order to alert their nestmates to danger. The acetates broadcast by the slave makers are so much stronger, however, that they have a long-lasting disruptive effect. For this reason Regnier and I named them "propaganda substances."

We believe we have explained an odd fact first noted by Pierre Huber 165 years ago in his pioneering study of the European slave-making ant *Formica sanguinea*. He found that when a colony was attacked by these slave makers, the survivors of the attacked colony were reluctant to stay in the same neighborhood even when suitable alternative nest sites were scarce. Huber observed that the "ants never return to their besieged capital, even when the oppressors have retired to their own garrison; perhaps they realize that they could never remain there in safety, being continually liable to the attacks of their unwelcome visitors."

Regnier and I were further able to gain a strong clue to the initial organization of slave-making raids. We had made a guess, based on knowledge of the foraging techniques of other kinds of ants, that scout workers direct their nest-mates to newly discovered slave colonies by means of odor trails laid from the target back to the home nest. In order to test this hypothesis we made extracts of the bodies of *F. subintegra* and of *Formica rubicunda,* a second species that conducts frequent, well-organized raids through much of the summer. Then at the time of day when raids are normally made we laid artificial odor trails, using a narrow paintbrush dipped in the extracts we had obtained from the ants. The trails were traced from the entrances of the nest to arbitrarily selected points one or two meters away.

The results were dramatic. Many of the slave-making workers rushed forth, ran the length of the trails and then milled around in confusion at the end. When we placed portions of colonies of the slave species *Formica subsericea* at the end of some of the trails, the slave makers proceeded to conduct the raid in a manner that was apparently the same in every respect as the raids initiated by trails laid by their own scouts. Studies of the slave-making species *Polyergus lucidus* and *Harpagoxenus americanus* by Mary Talbot and her colleagues at Lindenwood College provide independent evidence that raids are organized by the laying of odor trails to target nests; indeed, this form of communication may be widespread among slave-making ants.

The evolution of social parasitism in ants works like a ratchet, allowing a species to slip further down in parasitic dependence but not back up toward its original free-living existence. An example of nearly complete behavioral degeneration is found in one species of the genus *Strongylognathus,* which is found in Asia and Europe. Most species in this genus conduct aggressive slave-making raids. They have characteristic saber-shaped mandibles for killing other ants. The species *Strongylognathus testaceus,* however, has lost its warrior habits. Although these ants still have the distinctive mandibles of their genus, they do not conduct slave-making raids. Instead an *S. testaceus* queen moves into the nest of a slave-ant species and lives alongside the queen of the slave species. Each queen lays eggs that develop into workers, but the *S. testaceus* offspring do no work. They are fed by workers of the slave species. We do not know how the union of the two queens is formed in the first place, but it is likely that the parasitic queen simply induces the host colony to adopt her after her solitary dispersal flight from the nest of her birth.

Thus *S. testaceus* is no longer a real slave maker. It has become an advanced social parasite of a kind

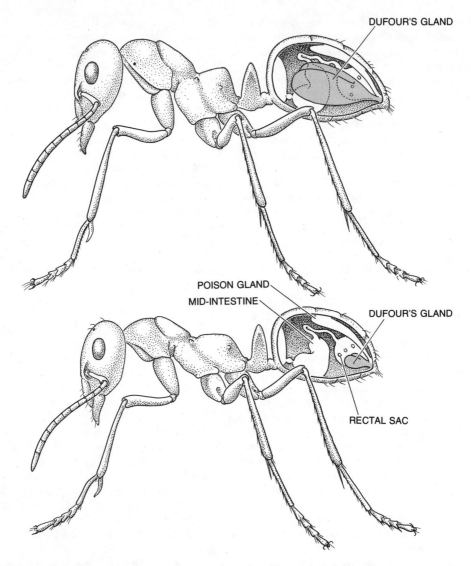

Figure 67 DUFOUR'S GLAND, which produces substances that serve as communication scents among ants, is much larger in the slave-making species *Formica subintegra* (top) than in the slave species *F. subsericea* (bottom).

The *subsericea* ant releases its scent to alert its nestmates to the presence of danger. The *subintegra* sprays its secretions at resisting ants during slave raids. The secretions are so strong that they create panic in the colony being attacked.

that commonly infests other ant groups. For example, many species of ant play host to parasites such as beetles, wasps and flies, feeding them and sheltering them.

Does ant slavery hold any lesson for our own species? Probably not. Human slavery is an unstable social institution that runs strongly counter to the moral systems of the great majority of human societies. Ant slavery is a genetic adaptation found in particular species that cannot be judged to be more or less successful than their non-slave-making counterparts. The slave-making ants offer a clear and interesting case of behavioral evolution, but the analogies with human behavior are much too remote to allow us to find in them any moral or political lesson.

How the Trypanosome Changes Its Coat

The parasite, which deprives much of Africa of meat and milk, survives in the bloodstream by evading the immune system. Its trick is to switch on new genes encoding new surface antigens.

. . .

John E. Donelson and Mervyn J. Turner

February, 1985

The African trypanosome is a microscopic animal, a protozoan, that spends part of its life cycle as a parasite in the blood of human beings and other mammals. There it causes a fatal neurological disease, trypanosomiasis, whose final stage in humans is sleeping sickness. The disease is endemic in a vast region of Africa defined by the range of the tsetse fly, the intermediate host that carries the trypanosome from one mammalian host to another. About 50 million people are at direct risk of contracting the disease; some 20,000 new cases are reported every year and thousands of other cases no doubt go unreported. Even more important than the direct threat to human beings is the fact that domestic livestock are susceptible to trypanosomiasis. Between them the trypanosome and the tsetse fly make some four million square miles of Africa—an area larger than that of the U.S.—uninhabitable for most breeds of dairy and beef cattle. Having little or no access to meat or dairy products, most of the human population is malnourished and susceptible to other diseases.

The key to the trypanosome's success is its ability to circumvent the mammalian immune system. A mammal ordinarily defends itself against viruses, bacteria or protozoa such as trypanosomes by manufacturing specific antibodies directed against the antigens, or "nonself" molecules, it recognizes on the foreign organism's surface; the antibodies bind to the antigens and neutralize or kill the invading organism. Some antibody-producing cells persist in the bloodstream and provide lasting immunity. Such immunity can also be elicited by a vaccine that mimics a natural infection.

Neither the immune response to infection nor vaccination can protect against trypanosomal infection. Even though these parasites are continually exposed in the bloodstream to the mammalian immune system (unlike the malaria parasite, which spends most of its life cycle sequestered within cells), they have evolved a way to evade the host's defenses: they keep changing the antigen that constitutes their surface coat. By the time the immune system has made new antibodies to bind to new antigens, some of the trypanosomes have shed their coat and replaced it with yet another antigenically distinct one. The host's overworked immune system is unable to cope with the infection, and so the parasites proliferate.

The molecular basis of this remarkable antigenic

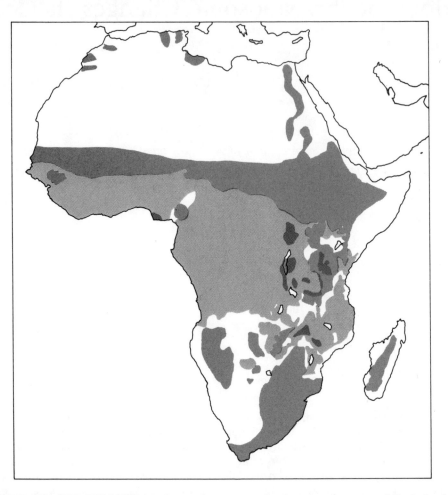

Figure 68 TSETSE FLY AND PARASITE together make some four million square miles of potential grazing land in Africa uninhabitable for most livestock. There is virtually no overlap between the range of the tsetse fly (*color*) and the cattle-raising regions of Africa (*gray*).

variation is under intensive study in a number of laboratories in Africa, Europe and the U.S. What has been learned suggests that there may be no way to help the immune system deal with trypanosomes once the parasites are established in the mammalian bloodstream, but that there may be other approaches to prevention or treatment.

The parasites that have evolved this very effective defense strategy are unicellular protozoans ranging from 15 to 30 thousandths of a millimeter in length. There are a number of species, which are classified on the basis of morphology and the hosts they infect. The two important species that infect human beings are *Trypanosoma rhodesiense* and *T.*

gambiense (named for the colonial territories where they were first identified). Three species are important for their infection of livestock: *T. congolense*, *T. vivax* and *T. brucei*. Because it is easy to grow in laboratory animals and cannot survive in human blood, *T. brucei* is the preferred species for investigation.

Trypanosomes, like many other parasites, assume different forms at different stages of their complex life cycle. The trypanosomes ingested by a tsetse fly along with an infected mammal's blood lodge in the fly's midgut, where they begin to undergo a series of biochemical and structural changes; in the process they lose their surface coat. After about three weeks they appear in the fly's salivary glands as the metacyclic form, which again carries a surface coat.

When the fly bites a mammal, metacyclic trypanosomes are introduced into the host's bloodstream, where they rapidly differentiate to a form that can proliferate. In humans the resulting disease can be either acute or chronic, depending on the infecting species. In both forms the disease first affects the blood vessels and lymph glands, causing intermittent fever, a rash and swelling. It is at this stage that the continuing battle with the host's immune system begins. Later the parasites invade the central nervous system; inflammation of the outer membranes of the brain leads to lethargy, coma and ultimately death.

The first indication that something keeps changing in the course of a trypanosome infection came early in this century. Physicians had observed a marked periodicity in the temperature of patients with trypanosomiasis. In 1910 the English investigators Ronald Ross and David Thomson, examining blood specimens taken from a patient every few days, noted that the changes in temperature were paralleled by a sharp rise and fall in the number of parasites in the blood. In reporting this finding Ross and Thomson quoted a suggestion, made by an Italian physician named A. Massaglia the year before, that "trypanolytic crises are due to the formation of anti-bodies in the blood. A few parasites escape destruction because they become used or habituated to the action of these antibodies. These are the parasites which cause the relapses." It was to be more than 50 years before this perceptive early insight could be confirmed and explained.

The explanation began to emerge in 1965, when Keith Vickerman of the University of Glasgow first described the thick surface coat covering the parasite's cell membrane. Soon it was discovered that individual trypanosome clones (populations descended from a single ancestral cell) have different surface coats. In 1968 Richard W. F. Le Page of the Medical Research Council's Molteno Institute in Cambridge, England, analyzed the antigenic surface proteins isolated from a number of clones. He found that each clone displayed a biochemically different protein; the differences were so marked that they

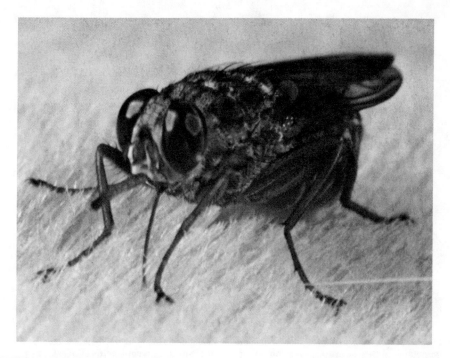

Figure 69 TSETSE FLY'S proboscis and distended abdomen are red with blood ingested from an experimental animal in this photograph made by Edgar D. Rowton of the Walter Reed Army Institute. The fly is the trypanosome's intermediate host and vector, in which the parasite goes through several developmental stages before it is injected into a mammalian host.

suggested each antigen must represent the expression of a different gene. (A gene is expressed by a cell when the DNA of the gene is transcribed to make a strand of messenger RNA, which is subsequently translated to make a protein.)

During the mid-1970's George A. M. Cross and his colleagues at the Molteno Institute generated four different clones from a trypanosome population infecting a laboratory animal. They showed that the surface coat consists of a matrix of identical glycoprotein molecules (proteins to which carbohydrate groups are attached) and that the glycoproteins are the same in all individuals in a single clone. Having determined the sequence of the amino acids (the subunits of protein chains) at the beginning of the glycoproteins of their four clones, they noted that the sequence was different in each case; the observation supported Le Page's proposal that dif-

ferent surface antigens must be encoded by different genes. These antigens are now called variable surface glycoproteins, or VSG's.

Early in the course of an infection the immune system generates antibodies shaped to bind to the particular VSG's it "sees" on the surface coat of invading parasites. The antibodies kill perhaps 99 percent of the original parasite population. A few individual trypanosomes escape, however, because they have turned on a different VSG gene and are covered by a new coat of VSG's to which the available antibodies cannot bind. These variant individuals give rise to a new population expressing the new set of VSG's; the new population grows while the immune system raises another set of antibodies, which eventually succeed in again killing some 99 percent of the parasites. By that time, however, a few of the parasites have changed their coat again,

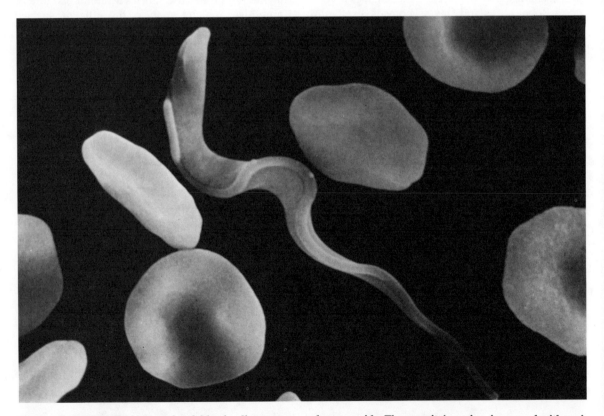

Figure 70 TRYPANOSOME and red blood cells are enlarged 5,500 diameters in this scanning electron micrograph made by Steven T. Brentano of the University of Iowa. The parasite, which is introduced into a mammal's bloodstream by the bite of the tsetse fly, is a unicellular animal (a protozoan) with a single flagellum extending along one side. The parasite's surface is covered with variable surface glycoproteins (VSG's). They are antigenic: the mammal's immune system makes antibodies that bind to them, killing the parasite. A trypanosome can change its surface coat, however, giving rise to a new parasite population that evades the host's antibody defense.

and therefore a new population proliferates. So it goes until the host dies. The available evidence suggests that the switch from one VSG to another takes place spontaneously. The host's immune system does not induce the switch but rather selects (by its inability to deal quickly with a particular new antigen) a variant that initiates a new population.

The total potential VSG repertory of a trypanosome is not known. In controlled experiments trypanosomes derived from a single parent cell have generated more than 100 distinct VSG's, giving no indication that their complement of VSG genes has been exhausted. It has recently been estimated that a single organism has a least a few hundred and perhaps as many as 1,000 or so VSG genes. (This means that from 5 to 10 percent of the parasite's total genetic capacity is devoted to antigenic variation.) In the wild the combined gene pool of all trypanosomes probably supplies genetic information to generate a virtually infinite number of antigenically distinct VSG's.

What is the structure of the VSG and how is it attached to the cell membrane? How does the parasite manage to express one VSG gene at a time (and only one) out of a repertory of hundreds of such genes? In order to learn something about the VSG's and about how they are displayed sequentially by the trypanosome, a number of research groups have been exploiting procedures based on recombinant-DNA technology.

The first step is to isolate the messenger RNA (mRNA) from a trypanosome clone. The mRNA is copied to make what is called complementary DNA (cDNA), which is combined with a carrier DNA and introduced into bacteria. By manipulating

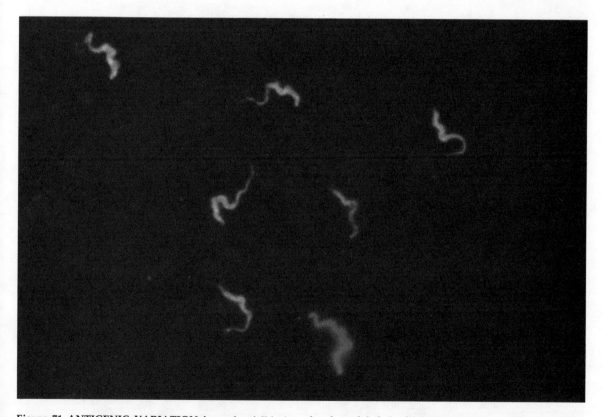

Figure 71 ANTIGENIC VARIATION is made visible in this double-exposure fluorescence micrograph made by Klaus M. Esser of the Walter Reed Army Institute of Research. Most of the trypanosomes are green because the antibody that recognizes and binds to their surface VSG has been labeled with a dye that gives off a green glow under ultraviolet radiation. One parasite is red. It has changed its surface coat and carries a different VSG, which is recognized by a different antibody: one that is linked to a red fluorescent dye.

the bacteria harboring the recombinant DNA, it is possible eventually to isolate a cDNA molecule that is in effect a copy of the VSG gene the trypanosome clone is expressing. By determining the sequence of nucleotides (the components of DNA) in the cDNA and translating it according to the genetic code, it is possible to predict the amino acid sequence of the VSG the gene encodes.

Partial or complete nucleotide sequences have been determined for some 15 cDNA's, and analysis of the deduced amino acid sequences reveals that each newly synthesized VSG is made up of about 500 amino acids. The first 20 or 30 of them at the *N* terminal of the protein constitute a signal peptide (a short protein chain) whose function is to help move the nascent VSG across the trypanosome's cell membrane; comparison of the cDNA-predicted amino acid sequence with the actual sequence of a few VSG's has shown that in the process the signal peptide is cleaved off. The sequence of the next 360 amino acids is quite different in most VSG's, and this variable region is presumably responsible for the parasite's antigenic diversity. The last 120 amino acids (at what is called the *C*-terminal end of the protein) are quite similar in various VSG's; on the basis of the degree of similarity in these homology regions, VSG's can be classified into two homology groups.

Ordinarily it is peptide sequences in the *C*-terminal region of a surface protein that anchor the protein in the cell membrane, but the VSG is different. The last 20-odd amino acids of the region are clipped off in the mature VSG and are replaced by a structure containing an unusual oligosaccharide (a

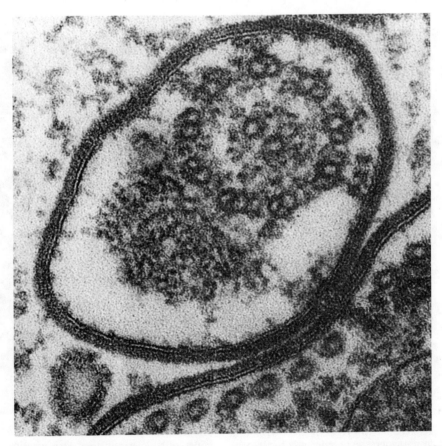

Figure 72 TRYPANOSOME'S SURFACE COAT of VSG's is visible as a diffuse, dark layer in an electron micrograph made by Laurence Tetley and Keith Vickerman of the University of Glasgow. A cross section of the parasite's body and flagellum has been enlarged about 190,000 diameters. The double membrane just inside the surface coat is the cell membrane.

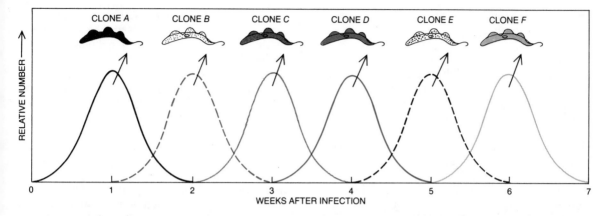

Figure 73 SUCCESSIVE WAVES of parasite proliferation in the blood are characteristic of trypanosomiasis. They result from antigenic variation. A population of parasites, some of which carry a particular antigen, VSG *A*, on their surface, proliferates in the bloodstream for a few days. The immune system raises antibodies against the population's antigens, killing most of the parasites. A few individual parasites survive by expressing new VSG genes that direct the synthesis of new antigens, such as VSG *B*; these parasites give rise to a new population that grows until the immune system manages to raise new antibodies against the new antigens. The cycle is repeated many times in the course of a chronic infection as parasites keep expressing new genes and displaying new VSG's. From each successive population it is possible to isolate individual trypanosomes and from them to grow clones expressing particular VSG's.

complex sugar molecule) that has a role in the anchoring. It seems to be very similar in all VSG's, regardless of their variable-region sequence, because antibodies raised against one oligosaccharide bind to all VSG molecules. Why does this cross-reacting determinant, as it is called, not induce natural immunity and why can it not serve as the basis of a vaccine? The reason is essentially that VSG's are packed into the surface coat in such a way that the immune system is confronted by the variable domain of the protein; the homology domain, including the cross-reacting determinant, is not exposed.

The cross-reacting determinant appears to be part of a larger oligosaccharide molecule, which is linked in turn to a phosphoglyceride carrying two fatty acid chains. It is the fatty acids that penetrate the cell membrane and hold the VSG in place. Why would the trypanosome substitute this complicated structure for the usual *C*-terminal cell-membrane anchor? The reason may be that the ability to release its surface coat rapidly is of central importance to the parasite. The trypanosome has an enzyme that can cleave the link to the fatty acids, thereby releasing the VSG from the membrane. Having the same specialized anchoring molecule, subject to cleavage by the same enzyme, on all VSG's regardless of their exact sequence provides a rapid and

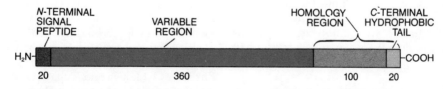

Figure 74 PROTEIN CHAIN of a typical VSG is composed of about 500 amino acids. The first 20 or so of these, at what is called the *N* terminal of the protein, constitute a signal peptide, which is cleaved from the protein before the VSG is implanted in the cell membrane. The next 360 amino acids (*color*) constitute the variable region, which is different in each antigenically distinct VSG. The final 120 amino acids at the *C* terminal are quite similar in each of two "homology groups" of VSG's. The last 20 amino acids of this region are cleaved from the chain and replaced by a large molecule that anchors the VSG in the cell membrane.

universal mechanism for stripping off one coat and substituting another.

Although comparison of cDNA sequences has revealed extensive differences among VSG's, it has become clear that a very few amino acid changes can suffice to generate antigenically distinct VSG's. Presumably these changes take place at particular antigenic sites within the 360-amino-acid variable region. To find these sites it will be necessary to know the three-dimensional structure of the variable region in detail. Some progress toward that goal has come recently from X-ray-crystallographic studies done by Don C. Wiley, Douglas M. Freymann and Peter Metcalf of Harvard University, in collaboration with one of us (Turner) at the Molteno Institute.

Five VSG's have been crystallized to date, but most progress has been made in determining the structure of the variable domain prepared from one of them. The resolution so far attained is enough to reveal the parts of a protein chain that are folded into the cylinder-like conformation known as an alpha helix. About half of the variable-region turns out to be in that form.

The variable region crystallizes as a dimer (a double molecule made up of two monomers), and indeed VSG's seem to aggregate as dimers in the surface coat of a living trypanosome as well. The dimer (or at least the resolved half of it) is shown by crystallography to be a bundle of alpha helixes. The core of the bundle is made up of two hairpin-shaped structures, one from each of the component monomers. At one end of the dimer the core interacts with two more helixes to form a six-helix bundle; at the other end the monomers diverge somewhat to form a distinctive head. This highly symmetrical structure must form the framework for the remaining half of the variable-domain sequence, unrevealed so far because it is not an alpha helix. We still do not know how the framework structure is oriented in the membrane (that is, which end is "out"), and so we cannot begin to guess where on the structure the major antigenic sites will be found.

Just what goes on when the trypanosome turns on one VSG gene after another, always synthesizing only one antigenically distinct surface glycoprotein at a time? Experiments designed to answer that question depend again on cDNA's of the kind described above. Now each such cDNA, which is in effect an artificial VSG gene, serves as a probe with which to locate copies of the same gene wherever they may be in the trypanosome genome (the total complement of genetic material).

The total genomic DNA is digested with a restriction enzyme, which cleaves the DNA at a specific site within a particular sequence of nucleotides. The result is that the genomic DNA is broken down into a large number of small fragments, each one slightly different in size. Having been separated according to their size by a process called gel electrophoresis, the fragments are transferred to nitrocellulose paper, to which they bind tightly. A cDNA representing a VSG gene, labeled with a radioactive isotope, is applied to the paper. The cDNA hybridizes with (binds to) any similar sequences it finds among the restriction fragments on the paper. Unbound cDNA is washed away and autoradiography reveals the sites of hybridization. One can thus determine how many different copies there are in a given trypanosome clone of the gene represented by the cDNA probe and whether or not the gene is in a different part of the genome in different clones.

In this manner Piet Borst and his colleagues at the University of Amsterdam and Cross and his associates at the Wellcome Research Laboratories in England demonstrated that when some VSG genes are expressed, an extra copy of the gene is present in the genome. They called it an expression-linked copy. Étienne Pays and Maurice Steinert of the Free University of Brussels went on to show that the mRNA for the expressed VSG gene is actually transcribed from this copy rather than from the original (basic-copy) gene that gave rise to it. Further analysis revealed that an expression-linked copy is always at a particular kind of site: near a telomere (the end of a chromosome). In other words, a single VSG gene out of the total repertory of such genes is expressed when it is duplicated and translocated to an expression site near a telomere; the switch from one VSG to another is often effected by the removal and degradation of one such copy and its replacement by the copy of another gene.

This copy-and-translocate mechanism is not the only source of antigenic variation. John R. Young, Phelix A. O. Majiwa and Richard O. Williams of the International Laboratory for Research on Animal Diseases in Nairobi found another mechanism. They discovered that in some cases the number of different fragments to which a particular cDNA probe hybridizes does not change when the VSG gene represented by the probe is expressed: there is

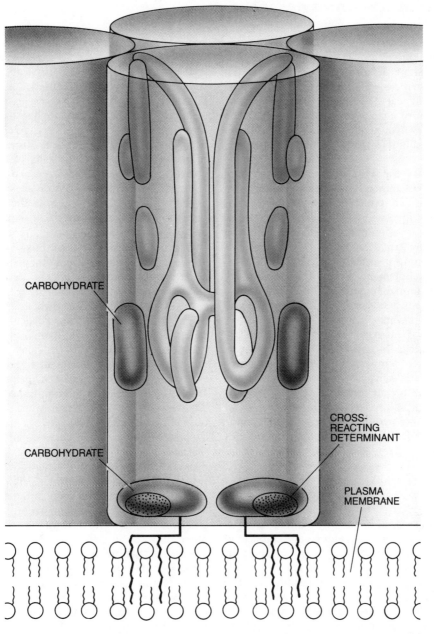

CARBOHYDRATE

CARBOHYDRATE

CROSS-REACTING DETERMINANT

PLASMA MEMBRANE

Figure 75 VSG'S OF SURFACE COAT may be assembled as indicated in this somewhat speculative drawing. The structure of part of the variable region of one VSG (*color*) is based on X-ray-crystallographic data; the full extent of the VSG and the location of its neighbors are indicated by the gray cylinders. The variable region is a dimer, or double molecule, that appears to be a bundle of the protein structures called alpha helixes. Carbohydrate molecules flank the bundle. Two more carbohydrates at the base of the VSG may incorporate a small sugar molecule: the cross-reacting determinant. Fatty acids extending from these two carbohydrates appear to anchor the VSG in the membrane.

no expression-linked copy. The probe hybridizes instead to a fragment that is different in size in each trypanosome clone, whether or not the clone is expressing the gene.

What this means is that some VSG genes are expressed without being duplicated and translocated. These genes turn out to be already at a site near a telomere. Their proximity to a telomere explains why their fragments vary in size. In trypanosomes and in some other organisms it is often the case that a short DNA sequence is repeated hundreds of times near a telomere. The number of these "tandem repeats" between a VSG gene and a telomere varies from clone to clone. As a result the fragment carrying a given telomere-linked VSG gene can be a different size in different clones.

About half of the VSG genes studied to date seem to be telomere-linked ones. Since there are hundreds of different VSG genes in the trypanosome genome, this suggests that there must be at least several hundred telomere-linked VSG genes in the genome, and so there must be hundreds of chromosomes in the parasite's nucleus! (The human chromosome complement is only 46.) Because the parasite has a normal amount of DNA for a unicellular animal, one would expect that some of the parasite's chromosomes must be very small, and indeed some of them are. A student of Borst's, Lex H. T. van der Ploeg, has applied a technique developed by David C. Schwartz and Charles R. Cantor of Columbia University to resolve trypanosomal nuclear DNA into four size classes: minichromosomes about 100,000 nucleotides long, other small chromosomes from two to seven times as long, middle-sized molecules (about two million nucleotides long) and others that are too long to be measured by this method.

Van der Ploeg found VSG genes on chromosomes in every size class. In one case he found a basic-copy gene on a large chromosome and its corresponding expression-linked copy on a middle-sized one, indicating that translocation of a duplicated molecule can take place between chromosomes. In another case a telomere-linked gene was not duplicated but was expressed at its normal site near the end of a large chromosome. These observations supported earlier evidence, accumulated in several laboratories, that there must be more than one site in the genome at which VSG genes can be activated; they are not all translocated to the same unique expression site near a particular telomere. Although proximity to a telomere is necessary for expression, it therefore cannot be sufficient. Other factors must come into play that select and activate a single VSG gene in a given organism for transcription into mRNA at a given time.

Some of the events attending this activation may contribute to increasing the diversity of VSG's. Pays and Steinert and their colleagues have reported cases in which a functional expression-linked copy was generated not by the duplication of a basic-copy gene but by the "recombination" of segments of at least two different telomere-linked genes, each of which codes for a part of the resulting VSG. If this kind of recombination is a general phenomenon, it must enable the trypanosome to make even more VSG's than its hundreds of genomic VSG genes can specify. It also suggests a reason for the telomeric location of expression sites: highly repetitive stretches of DNA such as the short tandem repeats near telomeres are particularly likely to undergo recombination.

Occasionally an expression-linked copy is not destroyed during the switch to the expression of another gene but instead "lingers" for a while in its expression site. At the University of Iowa one of us (Donelson) has recently shown that the sequences at the boundaries of one particular expression-linked copy are virtually identical with sequences bounding a telomere-linked gene. This, together with similar findings from other groups, suggests that some telomere-linked VSG genes may be previously expressed expression-linked copies that have survived the switch to a new gene. Perhaps two chromosomes exchange regions adjacent to their telomeres, so that an expression-linked copy is removed from its original expression site and placed near a different telomere, where it is available for future expression. Perhaps, on the other hand, a single small segment of DNA in the genome serves as a mobile control element: an enhancer of transcription that can move from one telomere to another and cause different VSG genes to be expressed. In that case an expression-linked copy gene could remain in its expression site but be turned off by the control element's departure.

There is another peculiarity of trypanosomes and related organisms. The mRNA's for VSG's (and for many, if not all, other proteins) always begin with the same specific sequence of 35 nucleotides. This sequence is not found in the corresponding gene or anywhere in the DNA surrounding the gene. Instead it is coded for by a repetitive DNA sequence separated from the gene. Small RNA molecules transcribed from this repeated DNA somehow provide copies of the 35-nucleotide sequence for the beginning of each mRNA. The 35-nucleotide sequence presumably has a role in the expression of trypanosome genes, but that role is not yet known.

L et us now summarize what is known about the mechanisms of antigenic variation in the trypanosome. There are hundreds of genes encoding VSG's in each organism's genome. They may be in the interior of chromosomes or near telomeres. Only one VSG is transcribed at a time, and that gene is always near a telomere. In order to be transcribed, an interior (basic-copy) gene needs to be duplicated and translocated (as an expression-linked copy) to one of many expression sites, all of which are near telomeres. A telomere-linked VSG gene, on the other hand, need not be duplicated in order to be expressed (although it may in fact sometimes be duplicated). Antigenic diversity may be further increased by recombination. The precise molecular mechanisms triggering the switch from one VSG

gene to another are still not known. It would appear, however, that the mechanisms are so complex and varied that it is almost impossible to circumvent them, and so the development of a vaccine against trypanosomes in the bloodstream is unlikely.

It may, however, be possible to develop a vaccine against metacyclic trypanosomes. As we mentioned above, the metacyclic stage is the final developmental stage in the tsetse fly's salivary glands, and it is metacyclic parasites that are injected into the bloodstream when a fly bites a mammal. Steven L. Hajduk, J. David Barry and Vickerman at Glasgow and Klaus M. Esser of the Walter Reed Army Institute of Research in Washington found that the metacyclic parasites can display on their surface only a reduced subset of VSG's—perhaps as few as 15. Collabor-

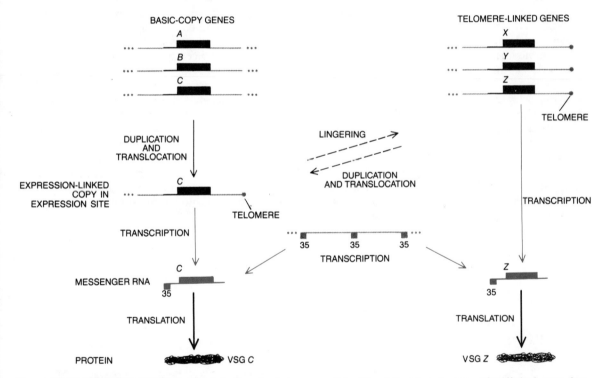

Figure 76 BASIC-COPY GENES in the interior of a trypanosome's chromosomes (*left*) are not ordinarily transcribed into messenger RNA (mRNA). For one of them (C) to be expressed it must be duplicated to provide an expression-linked copy. The DNA of the copy is translocated to an expression site near a telomere (the end of a chromosome) and is transcribed into mRNA, which is translated to make the VSG protein. Other VSG genes are already near a telomere (*right*). These telomere-linked genes can be ex-

pressed without forming an expression-linked copy, but occasionally they too are duplicated and translocated. An expression-linked copy is usually lost during an antigenic switch but sometimes "lingers" to become an unexpressed telomere-linked gene. Strangely, a 35-nucleotide sequence at one end of the mRNA is not encoded by the VSG gene but is specified instead by a separate region of repeated DNA.

ating with Esser, one of us (Donelson) has studied the cDNA's of several metacyclic VSG's and shown that the proteins have about the same C-terminal homology region as bloodstream VSG's; apparently they attach to the cell membrane in the same way. Moreover, the genes for the metacyclic VSG's seem to be telomere-linked ones, like many of the bloodstream VSG genes. It is therefore not yet clear why metacyclic parasites cannot express the full range of VSG's. There may nonetheless be a way to take advantage of their limited repertory or of their switching mechanism in order to make an effective vaccine.

The drugs currently available for treating trypanosomiasis are extremely toxic and also cannot prevent reinfection, but new forms of chemotherapy may yet be found. The trypanosome cannot survive in a mammalian host without its surface coat; a drug that interferes with the phosphoglyceride anchoring the VSG in the cell membrane, or that activates the enzyme releasing the VSG, might therefore be an effective therapeutic agent. Mammalian messenger RNA's do not have the unusual 35-nucleotide sequence described above; a drug that interferes with its synthesis might therefore selectively disable the parasite. Drugs might also be found that act against two subcellular organelles that seem to be unique to trypanosomes. One is the glycosome, a membrane-bounded aggregation of enzymes; the other is the kinetoplast, an appendage of the parasite's single large mitochondrion. A drug interfering with a metabolic function of these organelles, or with some other unique metabolic pathway as yet undiscov-ered, might kill the trypanosome without harming its mammalian host.

There are some other possible approaches. The tsetse fly can be eradicated briefly in small areas by spraying with insecticides or by the distribution of sterile male flies, but such methods cannot be effective for a region that extends for some four million square miles and more than a score of countries. A few breeds of cattle herded by nomadic tribes do seem to have developed partial resistance to trypanosomiasis. They do not yield much meat or milk, but it may be possible to cross them with more productive breeds. It is also conceivable that wild animals such as the eland or the oryx, which seem not to be harmed by trypanosomes, might be domesticated and take the place of cattle.

The past few years have made it clear that African trypanosomes are superb experimental organisms. Continued investigation of their variable surface coat will provide basic information about such diverse subjects as the control of gene expression, the attachment and functioning of membrane proteins, the structure and replication of chromosomal telomeres and the molecular mechanisms that generate biological diversity.

Fundamental knowledge of this kind should in turn contribute to a truly pressing public health task: the control of trypanosomiasis. The next few years should show if new basic information about trypanosomes can be applied to control and perhaps eventually eradicate trypanosomiasis.

Viroids

They are the smallest known agents of infectious disease: short strands of RNA.
They cause several plant diseases and possibly are implicated in enigmatic diseases
of man and other animals.

• • •

T. O. Diener

January, 1981

There are certain infectious diseases of plants that cannot be associated with any of the usual causative agents such as fungi, microorganisms or viruses. The only notable thing about a plant afflicted with one of these diseases seems to be that small molecules of an unusual form of the genetic material ribonucleic acid (RNA) can be isolated from its tissues. Such molecules cannot be detected in a healthy plant of the same species; if they are introduced into a healthy plant, they proliferate and give rise to the characteristic symptoms of the disease. In other words, these RNA molecules, which are called viroids, are the causative agent of the disease in question. They are the smallest known agents of infectious disease.

Viroids are much smaller than viruses and much simpler. Whereas a typical virus consists of genetic material (either RNA or DNA) surrounded by a protein coat, a viroid is nothing more than a very short strand of RNA. To date viroids have been identified with fewer than a dozen specific diseases, all of which affect higher plants, but there are indications that they may also cause animal diseases, perhaps including some rare nerve diseases that affect human beings.

In certain respects the discovery of viroids recalls the discovery of viruses some 75 years earlier. The conclusive demonstration by Louis Pasteur, Robert Koch and others that microorganisms, notably bacteria, were responsible for a number of infectious diseases gave rise by the 1890's to a general assumption that all infectious diseases must be caused by microorganisms. Investigations by the Russian microbiologist Dmitri Ivanovski and the Dutch botanist Martinus Beijerinck of a plant disease, tobacco mosaic, disproved that generalization by showing that the causative agent of the disease would pass through filters with pores small enough to retain bacteria. Their concept of a "filterable virus" finally led to the identification of virus particles and an understanding of their role as pathogens.

A new generalization took hold: all infectious diseases of plants and animals were widely held to be caused either by microorganisms or by viruses. Until quite recently any transmissible disease with which no microorganism or other agent could be

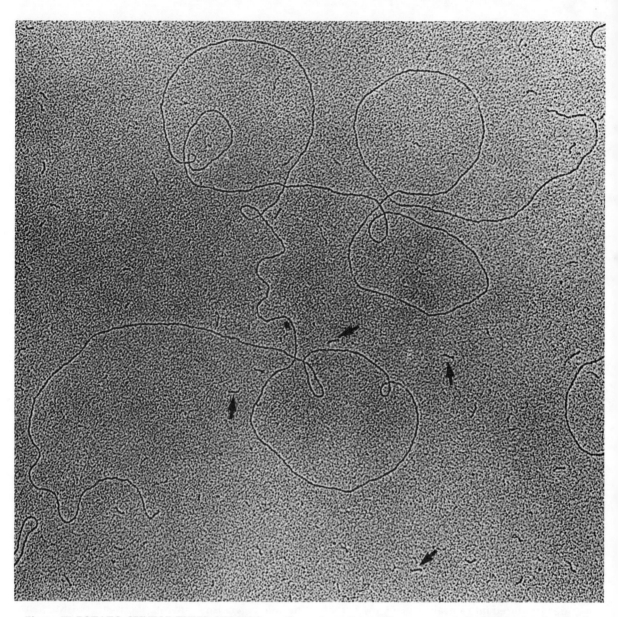

Figure 77 POTATO SPINDLE-TUBER VIROIDS are the scattered, short rods, each about 3.5 millimeters long, visible in this electron micrograph made by José M. Sogo and Theo Koller of the Swiss Federal Institute of Technology. They are short single strands of RNA, but in their "native" form, as here, they have a collapsed-circle or hairpin con- formation and hence appear to be double-strand molecules comparable in width to the long double-strand DNA of the bacteriophage (bacterial virus) T7, included to provide a size standard. The T7 DNA is about 14 micrometers (thousandths of a millimeter) long; the native viroid (*see arrows*) is only about 50 nanometers (.05 micrometer) long.

associated was automatically assumed to be caused by a virus; if the virus could not be isolated, it was assumed to be simply too elusive to be detected by the available techniques. It was the investigation of a plant disease that contravened this generalization too and led to the first isolation of a viroid.

The spindle-tuber disease of potatoes, which gives rise to gnarled, elongated tubers that are sometimes dissected by deep cracks in the surface, was first noted in the northeastern U.S. in the 1920's. Plant pathologists soon established that the disease was transmissible and that no bacteria or other microorganisms were consistently associated with it, and so it was assumed to be caused by a virus. Over the years several attempts were made to isolate the putative virus, always without success. That did not seem surprising; many viruses are quite hard to isolate and purify.

In 1962 William B. Raymer and Muriel J. O'Brien, working at the Agricultural Research Center of the U.S. Department of Agriculture in Beltsville, Md., found that the spindle-tuber agent could be transmitted from potato plants to young tomato plants, in which it multiplied and caused characteristic symptoms (stunt and twisted leaves) within about two weeks—much faster than in potatoes, where the disease usually becomes manifest only after the tubers have developed. That observation made it possible to do systematic experiments far more expeditiously. Raymer soon found he could prepare highly infectious extracts by grinding infected tomato leaves in a phosphate solution. It seemed that

purification of the virus should not present any great difficulty; one or two cycles of differential centrifugation should do the trick.

In that procedure an infectious extract is exposed alternately to moderate and very high centrifugal forces by spinning in a centrifuge. Centrifugation at about 10,000 times gravity tends to sediment impurities, whereas any virus particles remain suspended in the supernatant liquid. Exposing the supernatant to a force of $100,000g$ or more in a ultra centrifuge sediments most virus particles into a pellet. The pellet is resuspended; the suspension is again centrifuged to remove more impurities and again ultracentrifuged to pelletize the virus. When Raymer subjected his infectious tomato-leaf extracts to this procedure, he found that most of the infectious material remained in the supernatant liquid even after exposure to $100,000g$ for four hours. That is, the supernatant remained highly infectious and the material in the pellet was only slightly so.

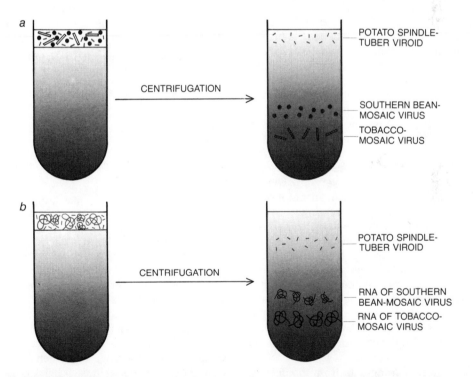

Figure 78 CENTRIFUGATION in a sucrose density gradient showed the agent of potato spindle-tuber disease had to be a small molecule of naked nucleic acid. In the centrifuge tube (a) a mixture of the viroid and two viruses is layered on a solution in which the sucrose concentration increases from the top to the bottom. Centrifugation for three hours at about 60,000 g separates the constituents by size into three bands, which are identified by infectivity assays; the PSTV remains at the top of the tube. When the viroid is centrifuged for 16 hours with the RNA of the two viruses (b), the viral RNA's move well down in the tube but the smaller viroid stays near the top.

There was something unusual about such a result, and when Raymer brought it to my attention, we decided to investigate the problem together.

One possible explanation—that the agent was a virus containing so much lipid (fat) that its density was particularly low—had already been ruled out: lipid solvents failed to inactivate the agent. The only other plausible explanation was that the infectious agent was remarkably small. To get a better estimate of its sedimentation properties we turned to centrifugation in a sucrose density gradient, a powerful technique developed by Myron K. Brakke of the University of Nebraska. We were surprised to find that the infectious agent sedimented not only at a lower rate than most virus particles we tested but also more slowly than the nucleic acid component of such particles. We had no choice but to suspect that the agent might consist of nothing but DNA or RNA. Treating the infectious material with the RNA-digesting enzyme ribonuclease inactivated the agent. Treatment with enzymes that break down DNA or proteins, on the other hand, had no effect on either the infectivity or the sedimentation behavior of the agent. This meant the essential element in infection had to be RNA and protein was probably not involved. Many other experiments confirmed these conclusions and convinced us that in living plants as well as in the laboratory the infectious RNA was not encapsulated, as it is in viruses, in a protein coat. In 1967 we proposed that the disease agent must be a free RNA.

Just how small was the infectious RNA molecule? It was hard to tell because the very low concentration of the RNA in infected tissue meant it could be detected only indirectly, by observing its biological activity: its ability to give rise to the characteristic symptoms when it was rubbed on the leaves of tomato plants. Even knowing the sedimentation rate of the RNA was not enough. Nucleic acid molecules of the same size can sediment at quite different rates depending on whether they are single-strand or double-strand molecules and whether they have a compact or a loose conformation. We were far from knowing such structural characteristics of the spindle-tuber agent.

It was only by combining data from the density-gradient sedimentation and from gel electrophoresis that I was able, with the help of Dennis R. Smith, to estimate the size of the spindle-tuber RNA. Gel electrophoresis exploits the fact that nucleic acid

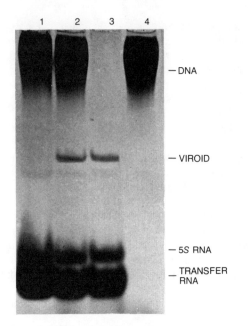

Figure 79 VIROID NUCLEIC ACID was eventually visualized and identified unequivocally as RNA by gel electrophoresis. Tomato-plant nucleic acids were placed in troughs at the top of a gel. Under the influence of an electric potential they migrated toward the positive pole (*bottom*) at a rate inversely proportional to the logarithm of their molecular weight and hence were separated into bands made visible by staining. Track No. 1 shows nucleic acids from healthy plants; nucleic acids from infected leaves were run in tracks Nos. 2, 3 and 4. The viroid is not seen in track No. 1. It is visible in track No. 2. It is not affected by the enzyme deoxyribonuclease, which digests the DNA (*track No. 3*), but it is itself digested, along with the other RNA, by the enzyme ribonuclease (*track No. 4*).

molecules, which are negatively charged, move toward the positive pole of an electric field. The preparation to be analyzed is placed in a well at one end of a polyacrylamide gel. When an electric current is applied, the various molecules are separated by size as they migrate through the gel at a rate inversely proportional to the logarithm of their molecular weight; eventually they form a series of bands each of which represents a collection of molecules of the same size. The bands can ordinarily be visualized by staining, and then the molecular weight of the material in each band is estimated by measuring its distance from the origin or from bands of identifiable molecules whose size is known. The small amount of spindle-tuber agent in our samples was not visible on staining, and so we depended again on its biological activity to determine its position in the gels. We cut each gel into thin slices and tested each slice for infectivity by rubbing it on the leaves of tomato plants.

The result was unequivocal: the infectious RNA was very small indeed. Our first estimate was that its molecular weight was about 50,000; more refined measurements eventually gave a value of about 130,000. This finding immediately raised a question that has yet to be answered: How could such a small RNA be a viable infectious agent? It is a well-established generalization in virology that most viruses have a genome, or total content of DNA or RNA, with a molecular weight of at least about a million. That much nucleic acid seems to be required for a virus to take over the genetic machinery of a host cell and subvert it to the cause of virus proliferation, and also to code for the necessary virus-specific proteins. A virus with less genetic information is termed "defective"; it cannot multiply on its own and must depend on genetic information provided by another virus—a "helper" virus—in the same cell.

We recognized that the potato spindle-tuber RNA did not have to code for the coat proteins characteristic of a virus, so that it could presumably make do with less genetic information than a typical virus can. A genome of the size we had estimated was only large enough, however, to code for protein with a total molecular weight of about 10,000, which is less than the size of a very small enzyme. The spindle-tuber RNA would have to depend largely, if not completely, on preexisting host enzymes for the synthesis of its progeny. This notion was difficult to accept, however, because plant cells were not thought to have any enzymes that could synthesize new RNA with the spindle-tuber RNA as a template. Such RNA-directed RNA polymerases had been identified only in cells infected by viruses and were apparently encoded by viral genes.

It seemed plausible, then, to regard the infectious RNA as a coatless defective virus that multiplied in plant cells with the aid of a helper virus. We therefore searched, in uninfected tomato plants, for any virus that might provide the necessary supplemental genetic information. All such efforts were fruitless. I finally became convinced that the infectious RNA was somehow able, in spite of its small size, to multiply independently of any helper virus. It was just possible, however, that the RNA we had detected represented not a single molecular species but

a population of different RNA molecules of about the same size, which could assemble to provide a more typical genome. We investigated this possibility by testing the effect of dilution and of ultraviolet irradiation on the infectivity of our RNA preparations; in both cases the slope of the infectivity curves seemed to rule out any possibility that several different molecules were involved. Evidently the agent of potato spindle-tuber disease differed drastically from viruses and was the first representative of a newly recognized class of subviral pathogens. In 1971 I proposed that such agents be called viroids.

By 1972 Joseph S. Semancik and Lewis G. Weathers of the University of California at Riverside showed that the agent of exocortis, an infectious disease of citrus trees previously assumed to be caused by a virus, was in fact a viroid. Then Roger H. Lawson of the Beltsville center found that the agent of a serious disease called chrysanthemum stunt had properties similar to those of the potato spindle-tuber viroid (PSTV). Collaboration between Lawson and me established that the agent is indeed a viroid.

At least five other plant diseases now appear to be caused by viroids. H. J. M. van Dorst and Dirk Peters of the Agricultural University at Wageningen in the Netherlands found that a viroid is responsible for cucumber pale-fruit disease. Charles P. Romaine and R. Kenneth Horst of Cornell University identified the agent of another chrysanthemum disease, chlorotic mottle, as a viroid; Matsuo Sasaki and Eishiro Shikata of Hokkaido University did the same for the agent of the stunt disease of hops. John W. Randles of the University of Adelaide found evidence that viroidlike RNA's are associated with cadang-cadang, a disease that has killed millions of coconut trees and caused large economic losses in the Philippines. Evidence that avocado sun-blotch disease may be caused by a viroid has been presented recently by N. A. Mohamed and Wayne Thomas of the New Zealand Department of Scientific and Industrial Research and by Peter Palukaitis and Robert H. Symons and their colleagues at the University of Adelaide. Undoubtedly viroids will be implicated as agents of still more plant diseases whose cause is not yet known.

Before the physical-chemical properties of viroids could be examined it was necessary to separate them from the nucleic acids of the cells they infect

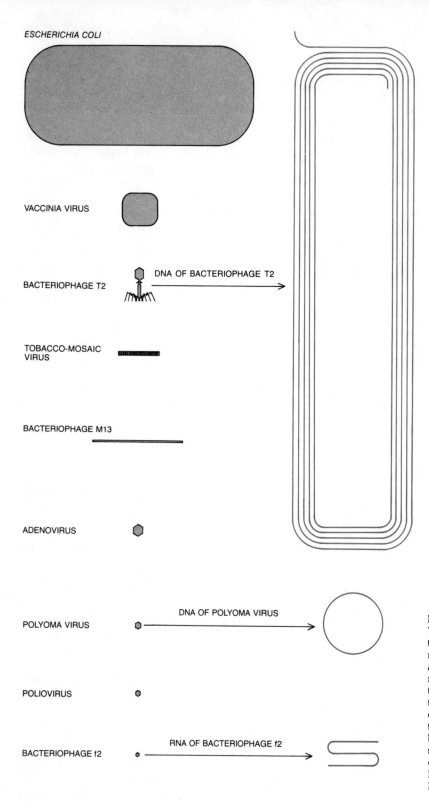

ESCHERICHIA COLI

VACCINIA VIRUS

BACTERIOPHAGE T2   DNA OF BACTERIOPHAGE T2

TOBACCO-MOSAIC
VIRUS

BACTERIOPHAGE M13

ADENOVIRUS

POLYOMA VIRUS   DNA OF POLYOMA VIRUS

POLIOVIRUS

BACTERIOPHAGE f2   RNA OF BACTERIOPHAGE f2

VIROID

Figure 80 SMALL SIZE OF A VIROID can be appreciated when the bacterium *Escherichia coli*, a number of viruses and the nucleic acid genomes of some of the viruses are compared with the native form of the potato spindle-tuber viroid (PSTV), all of them enlarged some 36,000 diameters in these drawings. The RNA of the bacterial virus f2, one of the smallest genomes that is large enough to direct independent replication of its virus, is much longer than viroid RNA.

and to purify them. The low concentration in plant tissues of the viroid RNA compared with host RNA made the separation difficult, and conventional purification methods applicable to protein-coated virus particles were of no help. By adopting improved methods of separating and purifying RNA and by working with large amounts of infected tissue we were eventually able to isolate viroids. We fractionated the small RNA's in extracts of healthy tomato leaves and in leaves infected by PSTV on electrophoresis gels and then measured the absorption of ultraviolet radiation by each RNA fraction. A prominent absorbance peak was present in the profile for infected leaves that was not present in the profile for healthy leaves; when the same fractions

were tested for their ability to infect healthy plants, the peak of the infectivity distribution was found to coincide precisely with the unique absorbance peak.

Taken together, these observations represented the first recognition of a viroid as a physical entity. As we succeeded in increasing the concentration of the viroid RNA it became possible to make the viroid fraction visible. It forms a distinct band on a gel treated with a dye that stains nucleic acids; the band is present only in infective extracts and is eliminated by the enzyme ribonuclease. To obtain PSTV essentially free of contaminating nucleic acids and in amounts sufficient for biophysical and biochemical analysis we proceed to cut the PSTV-containing slice out of a large number of gels, to extract the

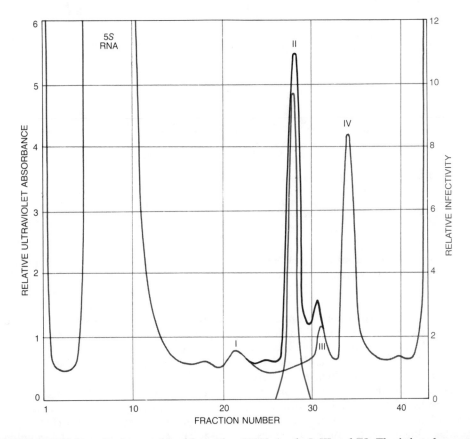

Figure 81 SMALL RNA'S from the leaves of healthy and of infected plants were separated according to size by gel electrophoresis and were then detected by the extent to which they absorbed ultraviolet radiation. Preparations from both healthy and infected leaves showed absorbance peaks for a ribosomal RNA (5S RNA) and three cellular RNA's (*peaks I, III and IV*). **The infected preparation also showed a unique fifth peak (II). Absence of peak II from healthy plants and its coincidence with the infectivity peak (*color*) indicated that it must represent the viroid RNA.**

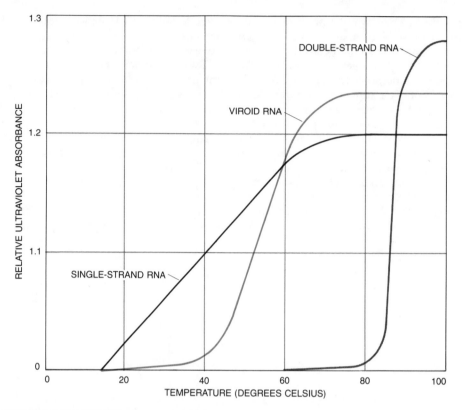

Figure 82 DENATURATION CURVES measure the change in an RNA's absorption of ultraviolet radiation as its base-paired regions separate when the temperature is increased. Here the characteristic curves for single-strand and dou-ble-strand RNA are compared with the curve for viroid RNA. Viroid is shown to be not double-strand RNA but rather an unusual single-strand RNA.

RNA with a solvent, to reconcentrate the RNA and to subject it to an additional electrophoresis cycle or two.

Once purified viroid preparations became available it became possible to investigate the infectious molecule's structure. To begin with, was it a single-strand RNA or a double-strand one? Both forms are found in viruses in which RNA constitutes the genome; cellular RNA's, which play various roles in the translation of genetic information from the cell's DNA genome into proteins, are almost invariably single-strand. Some early experiments (based, for example, on the rate at which a PSTV preparation's infectivity was reduced by treatment with ribonuclease) had suggested the viroid was a double-strand RNA, but contradictory results had been obtained by different analytical methods. If purified viroids could be seen clearly in

an electron micrograph, it was hoped, then simply measuring their width should resolve the ambiguity. The first successful micrographs, made from our preparations by José M. Sogo and Theo Koller, working at the Swiss Federal Institute of Technology in Zurich, showed a uniform population of rods with an average length of 50 nanometers (millionths of a millimeter), in good agreement with our earlier size estimate. The width of the viroids seemed to be similar to that of double-strand viral DNA seen in the same micrograph, suggesting the viroid too was double-strand. This turned out not to be the case, as we found by determining the denaturation properties of the viroid.

To briefly review the nature of DNA and RNA, a strand of nucleic acid is a chain composed of four subunits called nucleotides. Each nucleotide is characterized by a projecting chemical group called a base. In DNA the bases are adenine (*A*), guanine

(G), cytosine (C) and thymine (T); in RNA the first three bases are the same but thymine is replaced by uracil (U). The bases are complementary, so that in RNA A pairs with U and G pairs with C and sometimes with U. It is the pairing of complementary bases by hydrogen bonding that links two single strands to make a double-strand molecule. Base pairing can also take place in a single strand: one region of a strand can fold back on a complementary region, forming a hairpinlike loop. Regions linked by base pairing are "denatured," or separated, when a nucleic acid molecule is heated to break the hydrogen bonds between complementary bases, and the rate and the range of temperatures at which denaturation takes place vary with the structure of the molecule. The extent of thermal denaturation is most conveniently determined by measuring a nucleic acid preparation's ultraviolet absorbance, which increases with denaturation.

There is a characteristic curve relating ultraviolet absorbance to temperature for single-strand RNA and a different characteristic curve for double-strand RNA. The curve we derived for viroid RNA was not at all like the one for double-strand RNA; it was more like the curve for single-strand RNA but significantly steeper. We concluded that in their "native," or undenatured, state viorids are single-strand molecules folded into a hairpinlike configuration, with extensive regions of intrastrand base pairing. This explained why viroids in electron micrographs had appeared to be double-strand mole-

cules, and it also explained the contradictory results of the earlier experiments. A viroid is a single-strand RNA molecule with such extensive intrastrand pairing that it displays some of the characteristics of a double-strand RNA.

Soon two groups of investigators succeeded in making electron micrographs of completely denatured viroid molecules. William L. McClements and Paul J. Kaesberg of the University of Wisconsin reported that their PSTV micrographs show a mixture of two kinds of molecules: a majority of threadlike linear molecules roughly twice as long as the native rods and a minority of circular molecules whose circumference is about the same as the length of the linear ones. On the other hand, a group headed by Heinz L. Sänger of the University of Giessen in West Germany reported that viroids are all circular; Sänger and his colleagues think the rare linear molecules in their micrographs are artifacts of the purification process. The two groups agree, however, that the circular molecules are closed loops formed by covalent bonds of the nucleotide chain, not linear molecules whose two ends are linked simply by intrastrand base pairing between complementary regions at the two ends.

We undertook to find out if both the linear and the circular forms are present and are biologically significant in living plants. Robert A. Owens of my laboratory managed to separate denatured linear molecules from circular ones by subjecting purified, denatured PSTV to gel electrophoresis under condi-

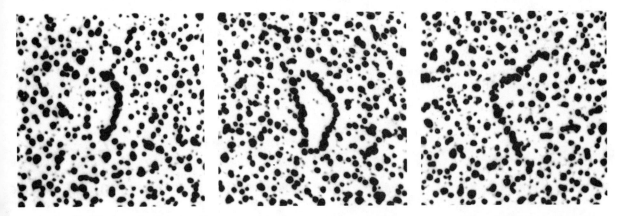

Figure 83 DENATURED VIROID MOLECULES seen in electron micrographs made by William L. McClements and Paul J. Kaesberg of the University of Wisconsin are enlarged here some 400,000 diameters. Native PSTV was denatured by formaldehyde treatment and prepared for microscopy under conditions largely preventing the reassociation of base pairs. Some PSTV molecules nonetheless reassumed their native structure (left). Of the denatured molecules, some were circular (middle), but most denatured molecules were linear (right).

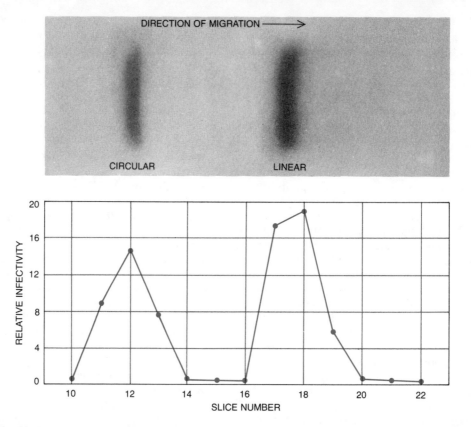

Figure 84 INFECTIVITY of circular and linear viroid molecules was established by denaturing PSTV and subjecting two samples to electrophoresis under conditions preventing renaturation. One gel was stained, revealing two PSTV bands (*top*); electron microscopy showed circular PSTV in one band, linear PSTV in the other. Corresponding slices in the other gel were ground up and tested for infectivity. Both circular and linear forms were shown to be infective (*bottom*).

tions that prevented renaturation; the linear molecules moved faster than the circular ones, so that the two forms ended up in two discrete bands. In collaboration with Russell L. Steere and Eric Erbe we were able to show that both forms were infectious. Then Ahmed Hadidi introduced a radioactive isotope of phosphorus into PSTV-infected tomato plants and monitored the incorporation of the isotope into RNA as the viroid replicated. After a short period of replication most of the radioactively labeled PSTV was in the circular form; as replication and isotope incorporation proceeded, an increasing number of the labeled molecules were found to be linear. This confirmed that both the circular and the

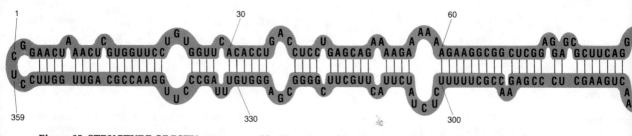

Figure 85 STRUCTURE OF PSTV was proposed by Hans J. Gross and his colleagues at the Max Planck Institute for Biochemistry in Munich, who worked out the 359-nucleotide sequence of the viroid and arranged the sequence to maximize base pairing. The vertical lines represent hydrogen bonds between permissible base pairs: $A-U$, $G-C$ and

linear forms are present in infected plants. It also suggested that the circular molecules may be precursors of the linear ones.

Meanwhile other workers were beginning to determine the chemical composition of viroids, first by the technique known as fingerprinting. An RNA is digested with an enzyme that cleaves the nucleotide chain only at particular sites and the resulting fragments of from one nucleotide to perhaps a dozen nucleotides are separated in two dimensions by electrophoresis and chromatography. The fragments form a two-dimensional pattern of spots that depends on the base sequence (the linear arrangement of the bases A, G, C and U) and hence is different for different molecules. Before fingerprinting was possible there had been some doubt about whether the viroids causing different diseases were in fact different molecules. Now Elizabeth Dickson and her colleagues at Rockefeller University showed the base sequence of PSTV and that of the citrus exocortis viroid were quite different from each other: the two viroids are distinct RNA species. As other viroids have been analyzed it has become clear that each is a distinct molecule with its own characteristic nucleotide sequence. On the other hand, different strains of a single viroid such as PSTV have sequences that differ from one another at only a few positions along the chain. Such results indicate that viroids are functional genetic systems whose characteristics, like those of other genetic elements, are determined by their nucleotide sequence.

In 1978 the complete primary structure—the entire nucleotide sequence—of PSTV was worked out by a group headed by Hans J. Gross of the Max Planck Institute for Biochemistry in Munich. The viroid is a chain of 359 nucleotides: 73 A's, 77 U's 101 G's and 108 C's. (The excess of G's and C's seems to be characteristic of viroids in general, judging from estimates of the base composition of several other viroid species.) Gross and his colleagues went on to propose a secondary structure, or conformation, for native PSTV by folding the primary sequence to allow the largest possible extent of base pairing and to account for other properties of the molecule, such as its resistance to cleavage by ribonucleases.

Their model is in good agreement with what had been predicted from physical observations: a closed single strand of RNA in which short double-strand regions (actually regions of intrastrand base pairing) alternate with still shorter single-strand regions (mismatched loops of unpaired bases). This is a novel and perhaps unique structure. Closed, circular single-strand RNA molecules have not previously been observed in nature. The high degree of intrastrand base pairing, which gives rise to the collapsed-circle or hairpin structure, is also most unusual. It is hard to believe these structural peculiarities do not have some connection with the biological activity of viroids.

The connection is not at all clear, however. Knowledge at the molecular level of the biological properties of viroids, and in particular how they replicate in a host cell and how they cause disease, is very limited. What is certain is that when a viroid is introduced into a host cell, it replicates without the assistance of a helper virus.

How does such a small RNA molecule induce its own synthesis in an infected cell? Might the viroid RNA encode the manufacture of a polypeptide (a short protein chain) that is essential for replication? Or is the viroid replicated solely by enzymes already present in a healthy plant? There is general agreement that viroids themselves are not translated into

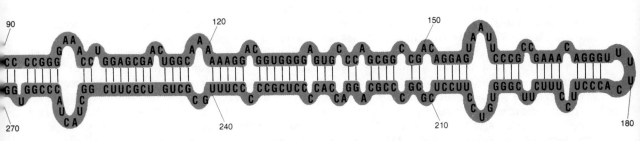

G – U; the colored band represents the covalently linked backbone of the molecule. Short base-paired regions alternate with shorter unpaired regions, resulting in structure resembling double-strand RNA.

CIRCULAR VIROID

LINEAR VIROID

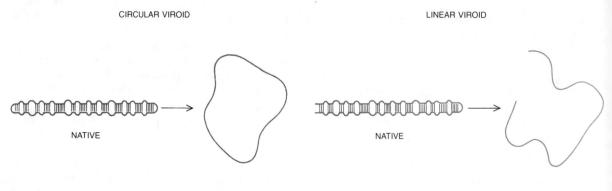

NATIVE

NATIVE

DENATURED

DENATURED

Figure 86 CLOSE SIMILARITY of the collapsed-circle structure of native circular molecules (*left*) and the hairpin structure of native linear ones (*right*) makes them impossible to distinguish by electrophoresis or in electron micrographs. When they are denatured, however, the difference is apparent, as is shown in these schematic drawings. Denatured circular molecules can be distinguished from denatured linear ones in electrophoresis gels (see Figure 84) and electron micrographs (see Figure 83).

protein, that is, they do not act as messenger RNA's encoding specific proteins. In laboratory preparations designed for the translation of RNA into protein they not only are inactive but also do not interfere with the translation of genuine messenger RNA's in the same preparation. Moreover, in plants infected with PSTV or the exocortis viroid no proteins can be found that are not also present in healthy plants. The synthesis of certain proteins is enhanced in infected plants, but Vicente Conejero and his colleagues at the Polytechnic University of Valencia have shown that those proteins are host proteins, not viroid-specific ones (at least in the case of plants infected by the exocortis viroid).

If viroids are not translated, they cannot encode enzymes. Presumably, then, their replication must rely entirely on enzyme systems of the host plant. Contrary to earlier assumptions, there are indeed enzymes in uninfected cells of a number of plant species, including the tomato, that can replicate RNA on an RNA template. These enzymes are obvious candidates for mediating the synthesis of new viroid RNA, but their involvement has yet to be demonstrated.

Whatever the source of the replicating enzymes, they can assemble new viroid strands only along a template: a nucleic acid having a sequence of bases complementary to that of the viroid. Is the template an RNA intermediary transcribed from the viroid or is it a DNA transcribed from the viroid by the en-

zyme called RNA-directed DNA polymerase? One ought to be able to find evidence for the presence of one or the other kind of template in infected cells. RNA sequences complementary to part of the citrus exocortis viroid can be found in infected cells but not in healthy ones, as was first shown by Larry K. Grill and Semancik. So far, however, it has not been shown that these complementary sequences represent the complete viroid, which presumably would be necessary to provide viable templates. On the other hand, we have identified some sequences complementary to PSTV in DNA from both healthy and PSTV-infected plants, but so far our findings have not been confirmed by other workers. In any case there has been no demonstration that such DNA sequences are involved in viroid replication.

This line of research depends primarily on hybridization experiments. Viroid RNA (or DNA with the same nucleotide sequence as the viroid), labeled so that it can be recognized, is mixed with an extract from infected plant tissue. Under conditions that promote base pairing the labeled nucleic acid probes "find" any complementary strands present in the plant extract and anneal to them to form hybrid double-strand molecules, which can be recognized by their labeling. In most of the investigations reported so far the probe has been a viroid isolated from an infected plant, purified to the extent possible and labeled. With such probes it is hard to exclude the possibility that any hybridization one observes may result from the base pairing of a host

RNA or DNA with some complementary plant RNA that still contaminates the viroid probe, rather than with the viroid itself.

Recently several investigators have been able to prepare probes by synthesizing strands of DNA that are complementary to a viroid. In my laboratory Owens and Dean E. Cress synthesized DNA complementary to almost the entire sequence of PSTV, made a second DNA strand complementary in turn to the first one, spliced the double-strand DNA into a plasmid (a small circle of bacterial DNA) and introduced the plasmid into the bacterium *Escherichia coli*, where it proliferated. This molecular cloning technique provides large quantities of an unambiguous probe, with which we hope soon to clarify the molecular mechanisms of viroid replication.

There is another fundamental question that has yet to be answered: How do viroids give rise to symptoms? And why do they cause disease only in certain susceptible plants, whereas they multiply in other plants without doing any discernible harm? Many of the symptoms are disturbances of growth and may therefore be the result of imbalanced growth-hormone activity. How viroids might bring that about is still an enigma. Viroids are found predominantly in the nuclei of infected cells, where genes are regulated—turned on or off. It is therefore at least possible that viroids function as abnormal regulatory molecules, somehow interfering with the control of genes encoding particular hormones.

S o far viroids have been clearly identified only in higher plants, but they (or very similar agents) may function in other forms of life as well. It seems reasonable to search for viroids wherever an infectious disease has been assumed to be caused by a virus but no virus particles have yet been identified.

That is the case for a group of brain diseases in animals known as the subacute spongiform encephalopathies. They include kuru and Creutzfeldt-Jakob disease in man, scrapie in sheep and goats and transmissible encephalopathy in mink. All four have been shown to be infectious nonbacterial diseases. They are commonly blamed on what are called "slow" viruses. Yet in spite of intensive efforts by many investigators no virus has been identified with any of them. Indeed, the causative agents have been shown to have properties that are either unknown or very rare in conventional viruses: great heat stability, extreme resistance to ultraviolet and ionizing radiations and unusual resistance to certain chemicals.

On the basis of comparisons between the properties of PSTV and the agent of scrapie I proposed in 1972 that the latter too may be a viroid. So far no clearly infectious nucleic acid has been isolated from the brain of animals with scrapie. Gordon D. Hunter and Richard H. Kimberlin and their colleagues at the Agricultural Research Council Institute for Research on Animal Diseases in England have, however, reported evidence for the presence of a small DNA molecule in infected tissue that is not present in the brain of healthy animals. Recently Richard F. Marsh and his colleagues at the University of Wisconsin have reported evidence that a component essential for scrapie infection can be inactivated with deoxyribonuclease, suggesting that the scrapie agent may be a coatless DNA molecule. Their findings have yet to be confirmed, and further characterization of the DNA will be required. It begins to appear, however, that viroidlike nucleic acids may be responsible for infectious diseases not only in plants but also in animals, perhaps including man.

The Authors

J. WILLIAM SCHOPF ("The Evolution of the Earliest Cells") is professor of paleobiology at the University of California at Los Angeles. He did his undergraduate work at Oberlin College and obtained his Ph.D. in biology from Harvard University in 1968, thereafter joining the U.C.L.A. faculty. Schopf's research on the earliest fossil records of life has involved fieldwork in North and South America, Australia, India, the U.S.S.R. and the People's Republic of China.

MARK A. S. McMENAMIN ("The Emergence of Animals") is assistant professor of geology at Mount Holyoke College. He did his undergraduate work in geology at Stanford University and then worked for five years for the U.S. Geological Survey. His Ph.D. is from the University of California at Santa Barbara. He has done geological fieldwork in a number of places, including Alaska and Sweden. McMenamin's principal avocational interests are computer programming, bio-dynamic-French intensive gardening, cross-country skiing, bicycle touring, hiking and "roughhousing with my children."

JAMES J. CHILDRESS, HORST FELBECK and **GEORGE N. SOMERO** ("Symbiosis in the Deep Sea") have worked together for several years exploring the adaptations by which animals thrive in the unique environment of the deep-sea vents. Childress is professor of biology at the University of California at Santa Barbara. He received his Ph.D. in biology from Stanford University and joined the faculty at Santa Barbara in 1969. Felbeck is assistant professor of biology at the Scripps Institution of Oceanography. He earned his doctorate from the University of Münster's Institute for Zoology in 1979 and did research at Scripps before joining the faculty. Somero, professor of biology in the marine biology research division at Scripps, received his Ph.D. from Stanford in 1967 and did postdoctoral studies at the University of British Columbia. He went to Scripps in 1970.

JOSEPH T. EASTMAN and **ARTHUR L. DEVRIES** ("Antarctic Fishes") are respectively associate professor of anatomy and zoology at Ohio University and professor of physiology at the University of Illinois at Urbana. Eastman did his undergraduate and graduate work at the University of Minnesota, obtaining his Ph.D. in 1970. He then taught at the University of Oklahoma Medical School for a year before he was appointed assistant professor of anatomical sciences there. He was on the faculty of the Brown University Medical School from 1973 until 1979, when he accepted his current position. Eastman has done field research in Antarctica on three occasions: once to study the anatomy of Antarctic seals and the other times to study the anatomy of Antarctic fishes. DeVries has a bachelor's degree from the University of Montana and a doctorate from Stanford University. It was while he was a graduate student that he first began his research on biological antifreeze compounds in coldwater fishes. From 1971 to 1976 he was associate research physiologist at the Scripps Institution of Oceanography. He then joined the faculty of the University of Illinois at Urbana-Champaign, being appointed full professor in 1984.

MICHAEL H. HORN and **ROBIN N. GIBSON** ("Intertidal Fishes") have done joint fieldwork in the Mediterranean and on the Atlantic coast of France. Horn is professor of biology at California State University at Fullerton. He earned a master's degree at the University of Oklahoma in 1965 and a Ph.D. from Harvard University in 1969 and went to Fullerton after spending a postdoctoral year at the Woods Hole Oceanographic Institution and the British Museum (Natural History). Gibson, principal scientific officer at the Scottish Marine Biological Association in Oban, obtained his Ph.D. in 1965 from the University of Wales and became interested in intertidal fishes as a postgraduate student there. He has continued his studies of the fishes both in Europe and in the Middle East, with occasional forays to the Pacific. Gibson was awarded a D.Sc. for contributions to the biology of shallow-water fishes.

BERND HEINRICH ("Thermoregulation in Winter Moths"), professor of zoology at the University of Vermont, has been studying thermoregulation in insects for almost 20 years. After receiving bachelor's and master's degrees from the University of Maine, he went to the University of California at Los Angeles, which granted his Ph.D. in 1970. He spent the next 10 years in the entomology department of the University of California at Berkeley before moving back east to Vermont. Heinrich is now examining the foraging habits of ravens but maintains that he has not changed fields: "I simply use different organisms to study the same questions."

ROBERT DEGABRIELE ("The Physiology of the Koala") teaches biology at Riverina College of Advanced Education, which is at the inland city of Wagga Wagga in New South Wales in Australia. He was educated at the University of New South Wales, receiving his M.Sc. in biology in 1977. His interest in marsupials and their evolution, he writes, "originated as a result of my being a student of T. J. Dawson, the author of the article titled 'Kangaroos' in the August 1977 issue of *Scientific American.*"

YOLANDE HESLOP-HARRISON ("Carnivorous Plants") is currently Leverhulme Research Fellow at the Welsh Plant Breeding Station of University College of Wales. She attended the University of Durham, where she obtained her doctoral degree in botany. She then lectured at the University of London and did research at the Royal Botanic Gardens in Kew. In addition to carnivorous plants her interests include the study of the lives of eminent Victorian scientists through their letters and other archival material and the archaeology of early technology in the English Midlands.

BERT HÖLLDOBLER ("Communication between Ants and Their Guests") is a professor of biology at Harvard University. He took up work on interspecific communication in social insects after obtaining his doctorate from the University of Würzburg in 1965. His scientific interests are in physiological and ecological aspects of insect behavior.

EDWARD O. WILSON ("Slavery in Ants") is professor of zoology at Harvard University. A graduate of the University of Alabama, Wilson acquired his Ph.D. in biology from Harvard in 1955; he has been a member of the Harvard faculty ever since. His research interests focus on the biology of the social insects, the classification of ants, sociobiology and biogeography, and he has published a number of books on these subjects, including *The Insect Societies* (1971) and *Sociobiology: The New Synthesis* (1975). His book *On Human Nature* won the Pulitzer Prize in 1979.

JOHN E. DONELSON and **MERVYN J. TURNER** ("How the Trypanosome Changes Its Coat") are respectively professor of biochemistry at the University of Iowa and director of the biochemical-parasitology program at the Merck Sharp & Dohme Research Laboratories in Rahway, N.J. Donelson holds a B.S. from Iowa State University and a Ph.D. in biochemistry from Cornell University. In 1974, after postdoctoral work at Cambridge and at Stanford University, he joined the faculty at the University of Iowa. His interest in tropical parasitic diseases originated during a stint in Ghana as a Peace Corps volunteer between college and graduate school; his involvement took him abroad again in 1980, when he traveled to Kenya as a visiting scientist at the International Laboratory for Research on Animal Diseases. Donelson is currently a Burroughs-Wellcome scholar in molecular parasitology. Turner is a graduate of the University of Sheffield, where he received a Ph.D. in organic chemistry in 1970. He pursued his interest in the biochemistry of membrane proteins as a postdoctoral fellow at Harvard University (1971 to 1974) and at Queen Victoria Hospital in London (1974 to 1977). In 1977 he became a fellow at the Medical Research Council's Molteno Institute at the University of Cambridge. Turner joined Merck Sharp & Dohme in 1985.

T. O. DIENER ("Viroids") is a research pathologist in the Department of Agriculture's Plant Virology Laboratory in Beltsville, Md. He received his undergraduate and graduate education at the Swiss Federal Institute of Technology in Zurich, where he was awarded his doctorate in plant pathology in 1948. He came to the U.S. in 1950 and spent most of the next decade studying virus-infected fruit trees at an experimental station of Washington State University. Since going to work for the Department of Agriculture in 1959, he has investigated various aspects of virus diseases in plants. It was in the course of this research that Diener discovered the novel group of disease agents now known as viroids.

Bibliographies

1. The Evolution of the Earliest Cells

Margulis, Lynn. 1970. *Origin of eukaryotic cells.* Yale University Press.

Carr, N. G., and B. A. Whitton, eds. 1973. *The biology of blue-green algae.* University of California Press.

Schopf, J. William. 1975. Precambrian paleobiology: Problems and perspectives. In *Annual Review of Earth and Planetary Sciences: Vol. 3,* eds. Fred A. Donath, Francis G. Stehli and George W. Wetherill. Annual Reviews Inc.

———. 1977. Biostratigraphic usefulness of stromatolitic Precambrian microbiotas: A preliminary analysis. *Precambrian Research* 5 (August): 143–173.

2. The Emergence of Animals

Brasier, M. F. 1979. The Cambrian radiation event. In *The origin of major invertebrate groups,* ed. M. R. House. Academic Press.

McMenamin, Mark A. S. 1982. A case for two late Proterozoic-earliest Cambrian faunal province loci. *Geology* 10 (June): 290–292.

Bengtson, Stefan, and Morris Simon Conway. 1984. A comparative study of lower Cambrian *Halkieria* and middle Cambrian *Wiwaxia*. *Lethaia* 17:307–329.

Glaessner, Martin. 1985. *The dawn of animal life: A biohistorical study.* Cambridge University Press.

Cowie, John W. 1985. Continuing work on the Precambrian-Cambrian boundary. *Episodes* 8 (June): 93–97.

3. Symbiosis in the Deep Sea

Janasch, Holger W., and Michael J. Mottl. 1985. Geomicrobiology of deep-sea hydrothermal vents. *Science* 229 (August 23): 717–725.

Jones, Meredith L., ed. 1985. Hydrothermal vents of the eastern Pacific: An overview. *Bulletin of the Biological Society of Washington* (December).

4. Antarctic Fishes

DeWitt, Hugh H. 1971. Coastal and deep-water benthic fishes of the Antarctic. In *Antarctic Map Folio Series, Folio 15,* ed. Victor C. Bushnell. American Geographical Society.

DeVries, Arthur L. 1982. Biological antifreeze agents in coldwater fishes. *Comparative Biochemistry and Physiology* 73A:627–640.

Eastman, J. T. 1985. The evolution of neutrally buoyant Antarctic fishes: Their specializations and potential interactions in the Antarctic marine food web. In *Antarctic nutrient cycles and food webs (Proceedings of the Fourth SCAR Symposium on Antarctic Biology),* eds. R. W. Siegfried, P. R. Condy and R. M. Laws. Springer-Verlag.

5. Intertidal Fishes

Gibson, R. N. 1982. Recent studies on the biology of intertidal fishes. *Oceanography and Marine Biology: An Annual Review* 20:363–414.

———. 1986. Intertidal fishes: Life in a fluctuating environment. In *The behavior of teleost fishes,* ed. Tony J. Pitcher. Johns Hopkins University Press.

Horn, Michael H., Margaret A. Neighbors and Steven N. Murray. 1986. Herbivore responses to a seasonally fluctuating food supply: Growth potential of two temperate intertidal fishes based on protein and energy assimilated from the macroalgal diets. *Journal of Experimental Marine Biology and Ecology* 103 (December 16): 217–234.

6. Thermoregulation in Winter Moths

Heinrich, Bernd. 1971. Temperature regulation of the sphinx moth, *Manduca sexta*: II, Regulation of heat loss by control of blood circulation. *Journal of Experimental Biology* 54 (February): 153–166.

Casey, Timothy M., and Barbara A. Joos. 1983. Morphometrics, conductance, thoracic temperature, and flight energetics of noctuid and geometrid moths. *Physiological Zoology* 56 (April): 160–173.

7. The Physiology of the Koala

Montgomery, G. G., ed. 1978. *The ecology of arboreal folivores.* Smithsonian Institution Press.

Bergin, T. J., ed. 1978. *The koala: Proceedings of the Taronga symposium on koala biology, management, and medicine.* Zoological Parks Board of N.S.W.

Degabriele, Robert, and T. J. Dawson. 1979. Metabolism and heat balance in an arboreal marsupial, the koala (*Phascolarctos cinereus*). *Journal of Comparative Physiology* 134:293–301.

8. Carnivorous Plants

Darwin, Charles. 1875. *Insectivorous plants.* Murray of London.

Heslop-Harrison, Yolande. 1975. Enzyme release in carnivorous plants. In *Lysosomes in biology and pathology: Vol. IV*, eds. J. T. Dingle and R. T. Dean. A. S. P. Biological and Medical Press.

Williams, Stephen E. 1976. Comparative sensory physiology of the *Droseraceae*—the evolution of a plant sensory system. *Proceedings of the American Philosophical Society* 120 (June 15): 187–204.

9. Communication between Ants and Their Guests

Seevers, Charles H. 1965. The systematics, evolution and zoogeography of staphylinid beetles associated with army ants (*Coleoptera, Staphylinidae*). *Fieldiana: Zoology* 47 (March 22): 139–351.

Akre, Roger D., and Carl W. Rettenmeyer. 1966. Behavior of *Staphylinidae* associated with army ants (*Formicidae: Ecitonini*). *Journal of the Kansas Entomological Society* 39 (October): 745–782.

Hölldobler, B. 1970. Contribution to the physiology of guest-host relations (myrmecophily) in ants, II: The relation between the imagos of *Atemeles pubicollis* and *Formica* and *Myrmica*. *Zeitschrift für Vergleichend Physiologie* 66:215–250.

10. Slavery in Ants

Talbot, Mary. 1967. Slave-raids of the ant *Polyergus lucidus* Mayr. *Psyche, Cambridge* 74:299–313.

Wilson, Edward O. 1971. *The insect societies*. Belknap Press of Harvard University Press.

Regnier, F. E., and E. O. Wilson. 1971. Chemical communication and "propaganda" in slave-maker ants. *Science* 172 (April 16): 267–269.

Wilson, E. O. 1975. *Leptothorax duloticus* and the beginnings of slavery in ants. *Evolution* 29:108–119.

11. How the Trypanosome Changes Its Coat

Desowitz, Robert S. 1981. New Guinea tapeworms and Jewish grandmothers: Tales of parasites and people. In *The fly that would be king*. W. W. Norton and Co.

Turner, M. J. 1982. Biochemistry of the variant surface glycoproteins of salivarian trypanosomes. *Advances in Parasitology* 21:69–153.

Ransford, Oliver. 1983. *Bid the sickness cease—disease in the history of black Africa*. John Murray, Ltd.

Freymann, D. M., P. Metcalf, M. Turner and D. C. Wiley. 1984. 6 Å-resolution x-ray structure of a variable surface glycoprotein from *Trypanosoma brucei*. *Nature* 311 (September 13): 167–169.

12. Viroids

Diener, T. O. 1979. Biology of viroids. In *Slow transmissible diseases of the nervous system*, eds. Stanley B. Prusiner and William J. Hadlow. Academic Press.

Sanger, H. L. 1979. Structure and function of viroids. In *ibid.*

Diener, T. O. 1979. *Viroids and viroid diseases*. John Wiley and Sons, Inc.

Gross, H. J., and D. Riesner. 1980. Viroids: A class of subviral pathogens. *Angewandte Chemie: International Edition in English* 19 (April): 231–243.

INDEX

Page numbers in *italics* indicate illustrations.